Common and Scientific
Names of Aquatic Invertebrates
from the United States and Canada:

Decapod Crustaceans

D1503526

Design of front cover and color plates by Beth D. McAleer

Publication of this book was made possible by a grant from the

Shell Companies Foundation

Travel and institutional support was extended to the Committee by the

National Marine Fisheries Service
National Oceanic and Atmospheric Administration
U.S. Department of Commerce

Additional travel support was provided by the

Fish and Wildlife Service
U.S. Department of the Interior

American Fisheries Society Special Publication 17

Common and Scientific Names of Aquatic Invertebrates from the United States and Canada:

Decapod Crustaceans

Austin B. Williams, *Chair*

Lawrence G. Abele, Darryl L. Felder, Horton H. Hobbs, Jr.,
Raymond B. Manning, Patsy A. McLaughlin, and
Isabel Pérez Farfante

*Committee on the Names of
Decapod Crustaceans of the
Crustacean Society*

Bethesda, Maryland
1989

The American Fisheries Society Special Publication series is a registered serial. A suggested citation format for this book follows.

Williams, A. B., L. G. Abele, D. L. Felder, H. H. Hobbs, Jr., R. B. Manning, P. A. McLaughlin, and I. Pérez Farfante. 1988. Common and scientific names of aquatic invertebrates from the United States and Canada: decapod crustaceans. American Fisheries Society Special Publication 17.

Library of Congress Catalog Number: 88-70618

ISBN 0-913235-49-0 (paper) ISSN 0097-0638

ISBN 0-913235-62-8 (cloth)

Address orders to

American Fisheries Society
5410 Grosvenor Lane, Suite 110
Bethesda, Maryland 20814, USA

CONTENTS

FOREWORD

The American Fisheries Society (AFS) Committee on Names of Aquatic Invertebrates was established by AFS President John J. Magnuson on September 30, 1981. The main goal of this Committee is to achieve uniformity and avoid confusion in vernacular nomenclature of aquatic invertebrates. The present charge by the Society to this Committee is as follows.

The Committee shall be responsible for studying and reporting on matters concerning common and scientific names of aquatic invertebrates, and shall prepare checklists of names to achieve uniformity and avoid confusion in nomenclature. The Chairman shall be custodian of the master checklists. The Committee shall coordinate its activities with those of other societies and organizations throughout the world. The Committee shall be composed of outstanding specialists in invertebrate taxonomy and nomenclature.

The Names of Aquatic Invertebrates Committee has benefited substantially from the long experience of and decisions reached by the AFS Names of Fishes Committee. The Names of Fishes Committee was originally appointed in 1933 as the result of a resolution adopted by the Society to form a permanent committee of experts in the field of ichthyology "to prepare and submit for publication a list of common names of fishes corresponding to the accepted scientific names." Because the AFS membership does not include much of the invertebrate taxonomic expertise needed to develop a comprehensive list of common and scientific names of aquatic invertebrates from the United States and Canada, it was decided to enlist the cooperation of other professional societies to accomplish the job.

The American Fisheries Society gratefully acknowledges the Committee on Names of Decapod Crustaceans of the Crustacean Society for developing this comprehensive list of common and scientific names of decapods. One CNAI member, Dr. Austin B. Williams, was particularly instrumental in bringing this list to fruition. The AFS Committee on Names of Aquatic Invertebrates has approved this list for publication.

Committee on Names of Aquatic Invertebrates

Donna D. Turgeon, *Chair*

Edward L. Bousfield	Kenneth L. Manuel
Thomas E. Bowman	Frank Maturo, Jr.
Stephen D. Cairns	David L. Pawson
Katharine Coates	Lynn B. Starnes
Larry L. Eng	Fred G. Thompson
Kristian Fauchald	Austin B. Williams
Mark Holliday	

LIST OF FAMILIES

INTRODUCTION

The purpose of this list is to provide a checklist of species and to recommend selected common names for North American decapod crustaceans; it is not to impose scientific names. Common names, we believe, can be stabilized by general agreement. Scientific names, on the other hand, cannot be fixed by democratic means; the limits of some taxonomic categories will inevitably shift with advancing knowledge. The Committee realizes that many of the users of this list are incompletely aware of the literature or are not interested in systematics. Therefore, the scientific nomenclature involved has been edited carefully with regard to spelling and authority; the nomenclature reflects the majority opinion of the Committee.

History and Background

In late 1981, the American Fisheries Society (AFS) created its Committee on Names of Aquatic Invertebrates (CNAI) to study and propose methods for standardizing common names of aquatic invertebrates in North America north of Mexico (Hutton 1985). Problems attend the increasing use of multiple vernacular names for many of these animals, including various decapod crustaceans. In daily affairs, such as marketing, government regulation, writing, and recreation, unambiguous familiar names, at least for many species, would help to reduce confusion in identity and thereby ease the flow of information. Although scientific names represent the basis for identity of such animals, they are seldom adopted in common use. Guidelines for selection of appropriate common names on which all users can agree are needed.

After deliberation, the CNAI decided that most of the principles of name selection developed by the long-standing AFS Names of Fishes Committee and set forth in AFS Special Publication 12 (Robins et al. 1980), could be modified to meet the special problems posed by aquatic invertebrates. Consensus was reached on a set of suggested principles for formation of common names for North American aquatic invertebrates, and the actual task of developing lists of these species was turned over to specialist groups. One such group was the Crustacean Society.

The Crustacean Society's Committee on the Names of Decapod Crustaceans has worked since 1983 to develop the first comprehensive list of North American decapod crustacean species. The Committee has endeavored to include common names for all species to which such names have been applied and for species with acknowledged economic, ecological, medical, recreational, or general interest; it has supplied names that it deems appropriate for other species as well. Some names have been selected from recently published species lists (Butler 1980; Holthuis 1980; Kessler 1985), but others have come from scattered sources. Many species have not been assigned a common name. The Committee feels that the rarity or obscurity of many such species renders common names unnecessary at this time. Should species in this category become objects of greater interest, common names then may be applied.

The Committee has drawn freely on introductory material developed for the AFS list of fish names (Robins et al. 1980) because that work is the model for this and other companion lists of aquatic invertebrates.

Area of Coverage

The present list purports to include all species of decapod crustaceans known from the fresh waters of North America north of Mexico, and those marine species known from within 320 km of the United States and Canada; coastal islands are embraced by this coverage, but not the West Indies. The species included are those whose occurrence in the region has either been authenticated in published accounts or verified in established research collections.

The list provides a general guide to distribution based on the following descriptors: A (Atlantic Ocean, including the Gulf of Mexico and the Arctic Ocean east of the Boothia Peninsula), P (Pacific Ocean, including the western Arctic Ocean), F (freshwater). A bracketed [I] following the indication of occurrence denotes an exotic or introduced species that has become established within the region.

The list contains 1,614 species: 509 shrimps; 364 lobsters, crayfishes, and lobsterlike forms; 285 hermit crabs, squat lobsters, porcelain, mole and sand crabs; and 456 brachyuran crabs. Of the marine species, 912 are distributed in contiguous waters of the Atlantic, 355 in the Pacific, and 36 occur in both regions. Freshwater species comprise 18 shrimps and 293 crayfishes, 289 of the latter in the Atlantic drainage, 4 in the Pacific, and 1 in both regions. At least three freshwater or estuarine species are exotic: an oriental crab and at least one South American shrimp were introduced to the Atlantic region and an oriental shrimp was introduced to the Pacific region. One Atlantic species, Harris mud crab, was introduced to the Pacific region. Species that are endangered are so designated in footnotes.

AFS Principles Governing Selection of Common Names

1. *A primary vernacular name shall be accepted for each species or taxonomic unit included.* Alternative published names may be listed in order of prominence. Rationale for selection of the primary name and etymologies may be indicated.

2. *No two species on the list shall have the same primary vernacular name.* Commonly used names of extralimital species should be avoided wherever possible.

3. *The expression 'common' as part of an invertebrate's name shall be avoided if possible.* Use of adjectives that also describe age or size and thus may have dual meanings shall be avoided as part of an invertebrate's name wherever possible (e.g., little, small, big, fat).

4. *Simplicity in names is favored.* Hyphens and suffixes shall be omitted (e.g., coonstriped shrimp) except where they are orthographically essential (e.g., speck-claw decorator crab), have a special meaning, or are necessary to avoid possible misunderstanding (e.g., red-ridged clinging crab). Compounded modifying words, including those that denote paired structures, usually should be treated as singular nouns in apposition with a group name (e.g., combclaw shrimp), but a plural modifier should usually be placed in adjectival form (e.g., banded coral shrimp), unless its plural nature is obvious (e.g., stalkeye sumo crab). Preference shall be given to names that are short and euphonious.

The compounding of brief, familiar words into a single name, written without a hyphen, may in some cases promote clarity and simplicity (e.g., longleg, sicklefoot, stalkeye), but the habitual practice of combining words, especially those that are lengthy, awkward, or unfamiliar is to be avoided. A guide for spelling of compound names is set forth in the latest edition of the *CEB Style Manual* (Council of Biology Editors, Bethesda, Maryland, USA).

5. *Common names shall not be capitalized in text use except for those elements that are proper names* (e.g., prickly lobsterette, but American lobster).

6. *Names intended to honor persons are discouraged in that they are without descriptive value.* In some large groups, identical specific patronymics in scientific names (sometimes honoring different persons) exist in related genera, and use of a patronymic in the common name is confusing. However, some patronymics that are already well established in literature, agency regulations, and industry (e.g., Tanner crab) may be retained. Apostrophes are not used.

7. *Only clearly defined and well-marked taxonomic entities* (usually species) *shall be assigned common names.* Most subspecies are not suitable subjects for common names, but those forms that are so different in appearance (not just in geographic distribution) as to be distinguished readily by lay people or for which a common name consitutes a significant aid in communication may merit separate names. There is a wide divergence of opinion concerning the criteria for recognition of subspecies. We have not named subspecies. Subspecies have importance in evolutionary inquiry but are rarely of significance to lay people or in those aspects of biological endeavor in which common names are of concern. The common name for the species should apply to all subspecies of a taxon and may be appropriately modified by those treating subspecies. The practice of adding geographic modifiers to designate regional populations makes for a cumbersome terminology.

Hybrids, cultured varieties, phases, and morphological variants are not named.

8. *The common name shall not be intimately tied to the scientific name.* Thus, the vagaries of scientific nomenclature (which are governed by the International Code of Zoological Nomenclature) do not entail constant changing of common names. The practice of applying a name to each genus, a modifying name for each species, and still another modifier for each subspecies, while appealing in its simplicity, has the defect of inflexibility. If an invertebrate is transferred from genus to genus, or shifted from species to subspecies or vice versa, the common name should nevertheless remain unaffected. It is not a primary function of

common names to indicate relationship. When two or more taxonomic groups (e.g., nominal species) are found to be identical, one name shall be adopted for the combined group.

This principle is regarded not only as fundamental to the achievement of stability, but as essential to the development of a true vernacular nomenclature.

9. *Names shall not violate the tenets of good taste.*

The foregoing principles are largely in the nature of procedural precepts. Those given below are criteria that are regarded as aids in the selection of suitable names.

10. *Colorful, romantic, fanciful, metaphorical, and otherwise distinctive and original names are especially appropriate.* Such terminology adds to the richness and breadth of the nomenclature and yields a harvest of satisfaction to the user. Examples of such names include yellowline arrow crab, bumblebee shrimp, visored shrimp, golden shrimp, and rusty grave digger.

11. *American Indian or other truly vernacular names are welcome for adoption as common names.* Names of Indian origin in current use include Chickamauga, Choctaw, and Chowanoke. In addition to aboriginal names, names of American decapods have been derived from people of non-English-speaking background, for example, gamba prawns (Italian).

12. *Commonly employed names adapted from traditional English usage* (e.g., crab, crayfish, prawn, shrimp) *are given considerable latitude in taxonomic placement.* Adherence to customary English practice is preferred if this does not conflict with the broad general use of another name. Many English names, however, have been applied to similar-appearing but often distantly related invertebrates in North America. We find shrimp in use for representatives of several decapod crustacean families. Crawfish or crayfish is applied to members of such diverse groups as the Cambaridae, Nephropidae, Palaemonidae, and Palinuridae. For widely known species, the Committee believes it preferable to recognize and adopt general use than to adopt bookish or pedantic substitutes. Thus, established practice should outweigh consistency with original English usage. This may not be well understood by some zoologists who may suggest strict adherence to the former usage.

13. *Structural attributes, color, and color pattern are desirable and are in common use in forming names.* Cerulean, flame, rough, squat, stellate, splendid, and a multitude of other descriptors decorate invertebrate names. Efforts should be made to select terms that are descriptively accurate, and to hold repetition of those most frequently employed (e.g., white, pink, spotted, banded) to a minimum.

Following tradition in American invertebrate zoology, we have attempted to restrict use of line or stripe to longitudinal marks that parallel the body axis and bar or band to vertical or transverse marks.

14. *Ecological characteristics are useful in making good names.* They too should be properly descriptive. Terms such as reef, pond, coral, sand, rock, riffle, freshwater, and mountain are well known in invertebrate names.

15. *Geographic distribution provides suitable adjectival modifiers.* Poorly descriptive or misleading geographic characterizations should be corrected unless they are too deeply entrenched in current usage. In the interest of brevity, it is usually possible to delete words such as lake, river, or ocean in the names of species (e.g., Obey crayfish, not Obey River crayfish).

16. *Generic names may be employed outright or in modified form (e.g., eualid from Eualus, argid from Argis) as common names.* Once adopted, such names should be maintained even if the generic name is changed. These vernaculars should be written in Roman and without capitalization. Brevity and euphony are of especial importance for names of this type.

17. *The duplication of common names of invertebrates and other organisms should be avoided if possible, but names in wide general use need not be rejected on this basis alone.* For example, the name hermit is employed for forest-dwelling tropical hummingbirds of the genus *Phaethornis* and in combination for various hermit crabs.

Plan of the List

The list is presented as a natural or phyletic sequence of suborders, infraorders, sections, superfamilies, and families of decapod crustaceans. The arrangement of these categories follows, in general, that adopted by Bowman and Abele (1982) in which authors and dates of establishment are given. Within families, the sequence is alphabetical by scientific name; the occasional attendant disadvantage of separating closely related forms within a family is regarded as more than offset by the ease in use.

Authors and dates of establishment for scientific names are included. These are commonly needed by persons who may not have ready access to the original references. Certain compound author names are hyphenated, but the compounds H. Milne Edwards and A. Milne Edwards are not hyphenated inasmuch as these authors themselves were inconsistent in use of the hyphen. Strict adherence to varying orthography of authorship for each of their species would be confusing to the uninitiated.

Use of the authority's name reflects current interpretation of the International Commission on Zoological Nomenclature (1985). In line with that code, the author's name follows the specific name directly and without punctuation if the species, when originally described, was assigned to the same genus in which it appears here; if the species was originally described in another genus, the author's name appears in parentheses. For example, Le Conte originally named the redjointed fiddler *Gelasimus minax*; it appears here as *Uca minax* (Le Conte, 1855).

Recommendations regarding the spelling of scientific patronyms are set forth in the International Code of Zoological Nomenclature (1985: Appendix D, III). Although recommendations regarding construction of patronyms have varied in the past, both the second and third editions of the Code have remained stable on these points and the Committee has elected to follow them.

Decisions have been made by majority opinion of the Committee. Thus, no committee member or outside counsel necessarily subscribes to all decisions reached. In many places, information available to the Committee exceeds that in the current literature, but the Committee has been cautious regarding inclusion of such information.

There are a great many vernacular names of North American decapod crustaceans, and extensive search would be necessary to assemble even a fraction of them. The Committee has attempted to reach consensus regarding a single set of accepted common names, realizing that in time the list will not only be more complete but may appropriately contain equivalent names recognized regionally or extraterritorially.

The synopsis of families that begins on page 1 serves as a Table of Contents. The main list appears on pages 9 to 47; the index begins on page 48.

Index

In this first comprehensive list of decapod crustacean names for North America, a single index incorporates the common and the scientific names. Page references are given for approved common names of species. Only one entry is given for the common name of a species; for example, the common name of *Callinectes sapidus* is entered only as "crab, blue." Page references are also given for the scientific names entered for family subdivisions, as well as for each genus and species. Each species is entered only under its specific name. For example, the scientific name of American lobster may be located at "*americanus, Homarus*" but not as "*Homarus americanus*," although the entry for *Homarus* guides the reader to the proper page. Scientific names that are synonyms of accepted scientific names are not included in this list.

Future of the Decapod Crustacean List

Subsequent to publication, proposed changes to the published decapod crustacean list will be collected, the list will be revised where necessary, and the second edition will be published 10 years after the first. Future editions are planned for every 10 years after the second edition.

Readers of the decapod crustacean list who wish to recommend changes in listed scientific or common names, reorganization of the phyletic arrangement, or additions or deletions of species should (1) clearly identify the desired change or changes, (2) briefly and specifically justify the change with reference to literature sources, and (3) send the request to the Chair, Committee on Names of Decapod Crustaceans, c/o American Fisheries Society, 5410 Grosvenor Lane, Suite 110, Bethesda, Maryland 20814, USA. All such requests will be considered by the Committee for the second edition.

Acknowledgments

The Crustacean Society's Committee on the Names of Decapod Crustaceans has had help from many contributors. Substantial counsel was given through several drafts of the list by David K. Camp, Fenner A. Chace, Jr., John S. Garth, Robert H. Gore, Janet Haig, David C. Judkins, Doyne W. Kessler, and Bradley G. Stevens. Others who helped with the preparation are M. James Allen, E. L. Bousfield, Richard C. Brusca, James T. Carlton, Deborah M. Dexter, Bruce E. Felgenhauer, Richard S. Fox, Richard W. Heard,

Jr., Diana R. Laubitz, William G. Lyons, Joel W. Martin, Paula M. Mikkelsen, Robert S. Otto, Mary K. Wicksten, and anonymous contributors. Composition of the scientific names was reviewed by George C. Steyskal, Systematic Entomology Laboratory, U.S. Department of Agriculture. The index was compiled and edited by Mark Holliday.

Members of the AFS Committee on Names of Aquatic Invertebrates from 1981 to 1986 with dates of appointment were: Robert F. Hutton, *Chair*, 1981; Larry L. Eng, 1982; Mark Holliday, 1982; Fred A. Mangum, 1985; Richard Neves, 1985; James R. Sedell, 1985; Scott E. Siddall, 1983; Lynn B. Starnes, 1982; Fred G. Thompson, 1983; Donna D. Turgeon, 1982; Elizabeth L. Wenner, 1982; Walter J. Wenzel, 1982; Austin B. Williams, 1982.

Travel funds for Committee members to attend work sessions, since 1981, have been provided, in large measure, by the National Marine Fisheries Service (National Oceanic and Atmospheric Administration, U.S. Department of Commerce) and the Fish and Wildlife Service (U.S. Department of the Interior) through cooperative agreements with AFS. We thank our home institutions for subsidizing our efforts on this project, for secretarial help, and for computer and duplicating facilities. Staff of the American Fisheries Society's international headquarters have helped in many ways, particularly Carl R. Sullivan, Executive Director, Robert L. Kendall, Managing Editor, and Sally M. Kendall, staff editor.

References

Bowman, T. E., and L. G. Abele. 1982. Classification of the recent Crustacea. Pages 1-127 *in* D. E. Bliss and L. G. Abele, editors. The biology of Crustacea, volume 1. Systematics, the fossil record, and biogeography. Academic Press, New York.

Butler, T. H. 1980. Shrimps of the Pacific coast of Canada. Canadian Bulletin of Fisheries and Aquatic Sciences 202.

Holthuis, L. B. 1980. Shrimps and prawns of the world. An annotated catalogue of species of interest to fisheries. FAO (Food and Agriculture Organization of the United Nations) Fisheries Synopsis 125.

Hutton, R. F. 1985. Names of aquatic invertebrates. (A name is a name is a name.) Fisheries (Bethesda) 10(6):16-20.

International Commission on Zoological Nomenclature. 1985. International code of zoological nomenclature, 3rd edition. International Trust for Zoological Nomenclature, London.

Kessler, D. W. 1985. Alaska's saltwater fishes and other sea life. Alaska Northwest Publishing, Anchorage.

Robins, C. R., R. M. Bailey, C. E. Bond, J. R. Brooker, E. A. Lachner, R. N. Lea, and W. B. Scott. 1980. A list of common and scientific names of fishes from the United States and Canada, 4th edition. American Fisheries Society Special Publication 12.

NAMES OF DECAPOD CRUSTACEANS

SCIENTIFIC NAME	OCCURRENCE[1]	COMMON NAME

PHYLUM, SUBPHYLUM, OR SUPERCLASS CRUSTACEA

CLASS MALACOSTRACA

SUBCLASS EUMALACOSTRACA

SUPERORDER EUCARIDA

ORDER DECAPODA

SUBORDER DENDROBRANCHIATA

Superfamily Penaeoidea—penaeoid shrimps

Solenoceridae—solenocerid shrimps

Hadropenaeus affinis (Bouvier, 1906)	A	....
Hadropenaeus modestus (Smith, 1885)	A	....
Hymenopenaeus aphoticus Burkenroad, 1936	A	....
Hymenopenaeus debilis Smith, 1882	A	....
Hymenopenaeus laevis (Bate, 1881)	A	....
Mesopenaeus tropicalis (Bouvier, 1905)	A	.... salmon shrimp
Pleoticus robustus (Smith, 1885)	A	.... royal red shrimp
Solenocera atlantidis Burkenroad, 1939	A	.... dwarf humpback shrimp
Solenocera mutator Burkenroad, 1938	P	....
Solenocera necopina Burkenroad, 1939	A	.... deepwater humpback shrimp
Solenocera vioscai Burkenroad, 1934	A	.... humpback shrimp

Benthesicymidae—benthesicymid shrimps

Bentheogennema borealis (Rathbun, 1902)	P	...... northern blunt-tail shrimp
Bentheogennema burkenroadi Krygier and Wasmer, 1975	P	.... Burkenroad blunt-tail shrimp
Bentheogennema intermedia (Bate, 1888)	A	....
Benthesicymus bartletti Smith, 1882	A	....
Benthesicymus brasiliensis Bate, 1881	A	....
Benthesicymus carinatus Smith, 1884	A	....
Benthesicymus cereus Burkenroad, 1936	A	....
Benthesicymus crenatus Bate, 1881	P	....
Benthesicymus iridescens Bate, 1881	A	....
Benthesicymus pectinatus (Schmitt, 1921)	P	....
Benthonectes filipes Smith, 1885	A	....
Gennadas bouvieri Kemp, 1909	A	....
Gennadas brevirostris Bouvier, 1905	A	....
Gennadas capensis Calman, 1925	A	....
Gennadas elegans (Smith, 1882)	A	....
Gennadas propinquus Rathbun, 1906	P	....
Gennadas scutatus Bouvier, 1906	A	....
Gennadas talismani Bouvier, 1906	A	....

---

[1]A = Atlantic; P = Pacific; F = freshwater; [I] = introduced and established.

SCIENTIFIC NAME	OCCURRENCE	COMMON NAME
Gennadas valens (Smith, 1884)	A	....

Aristeidae—gamba prawns

Aristaeomorpha foliacea (Risso, 1827)	A	.... giant gamba prawn
Aristeus antillensis		
A. Milne Edwards and Bouvier, 1909	A	...... purplehead gamba prawn
Hemipenaeus carpenteri Wood-Mason, 1891	A–P	....
Hemipenaeus spinidorsalis Bate, 1881	P	....
Hepomadus tener Smith, 1884	A	....
Plesiopenaeus armatus Bate, 1881	A	....
Plesiopenaeus coruscans (Wood-Mason, 1891)	A	....
Plesiopenaeus edwardsianus (Johnson, 1868)	A	.... scarlet gamba prawn

Penaeidae—penaeid shrimps

Funchalia villosa (Bouvier, 1905)	A	....
Metapenaeopsis gerardoi Pérez Farfante, 1971	A	....
Metapenaeopsis goodei (Smith, 1885)	A	.... velvet shrimp
Metapenaeopsis smithi (Schmitt, 1924)	A	....
Parapenaeus americanus Rathbun, 1901	A	....
Parapenaeus politus Smith, 1881	A	.... rose shrimp
Penaeopsis serrata Bate, 1881	A	.... pinkspeckled shrimp
Penaeus aztecus Ives, 1891	A	.... brown shrimp
Penaeus brasiliensis Latreille, 1817	A	.... pinkspotted shrimp
Penaeus californiensis Holmes, 1900	P	.... yellowleg shrimp
Penaeus duorarum Burkenroad, 1939	A	.... pink shrimp
Penaeus setiferus (Linnaeus, 1767)	A	.... white shrimp
Trachypenaeus constrictus (Stimpson, 1871)	A	.... roughneck shrimp
Trachypenaeus similis (Smith, 1885)	A	.... roughback shrimp
Xiphopenaeus kroyeri (Heller, 1862)	A	.... seabob

Sicyoniidae—rock shrimps

Sicyonia brevirostris Stimpson, 1871	A	.... brown rock shrimp
Sicyonia burkenroadi Cobb, 1971	A	.... spiny rock shrimp
Sicyonia dorsalis Kingsley, 1878	A	.... lesser rock shrimp
Sicyonia ingentis (Burkenroad, 1938)	P	.... ridgeback rock shrimp
Sicyonia laevigata Stimpson, 1871	A	....
Sicyonia olgae Pérez Farfante, 1980	A	....
Sicyonia parri (Burkenroad, 1934)	A	....
Sicyonia stimpsoni Bouvier, 1905	A	.... eyespot rock shrimp
Sicyonia typica (Boeck, 1864)	A	.... kinglet rock shrimp

Superfamily Sergestoidea

Sergestidae

Acetes americanus Ortmann, 1893	A	....
Lucifer faxoni Borradaile, 1915	A	....
Lucifer typus H. Milne Edwards, 1837	A	....
Petalidium suspiriosum Burkenroad, 1937	P	....
Sergestes arcticus Krøyer, 1855	A	....

SCIENTIFIC NAME	OCCURRENCE	COMMON NAME
Sergestes armatus Krøyer, 1855	A	
Sergestes atlanticus H. Milne Edwards, 1830	A–P	
Sergestes consobrinus Milne, 1968	P	
Sergestes corniculum Krøyer, 1855	A	
Sergestes cornutus Krøyer, 1855	A	
Sergestes edwardsii Krøyer, 1855	A	
Sergestes erectus Burkenroad, 1940	P	
Sergestes henseni (Ortmann, 1893)	A	
Sergestes orientalis Hansen, 1919	P	
Sergestes paraseminudus Crosnier and Forest, 1973	A	
Sergestes pectinatus Sund, 1920	A	
Sergestes pestafer Burkenroad, 1937	P	
Sergestes sargassi Ortmann, 1893	A	
Sergestes similis Hansen, 1903	P	
Sergestes vigilax Stimpson, 1860	A	
Sergia bigemmea (Burkenroad, 1940)	P	
Sergia bisulcata (Wood-Mason, 1891)	P	
Sergia creber (Burkenroad, 1940)	P	
Sergia fulgens (Hansen, 1919)	P	
Sergia gardineri (Kemp, 1913)	P	
Sergia grandis (Sund, 1920)	A	
Sergia inequalis (Burkenroad, 1940)	P	
Sergia japonica (Bate, 1881)	A	
Sergia robusta (Smith, 1882)	A	
Sergia scintillans (Burkenroad, 1940)	A	
Sergia splendens (Sund, 1920)	A	
Sergia tenuiremis (Krøyer, 1855)	A	

SUBORDER PLEOCYEMATA

INFRAORDER STENOPODIDEA

Stenopodidae

Microprosthema semilaeve (von Martens, 1872)	A	crimson coral shrimp
Odontozona libertae Gore, 1981	A	
Odontozona spongicola (Alcock and Anderson, 1899)	P	
Stenopus hispidus (Olivier, 1811)	A	banded coral shrimp
Stenopus scutellatus Rankin, 1898	A	golden coral shrimp

INFRAORDER CARIDEA

Superfamily Atyoidea

Atyidae

Palaemonias alabamae Smalley, 1961	F	Alabama cave shrimp
Palaemonias ganteri Hay, 1901	F	Mammoth Cave shrimp
Potimirim potimirim (Mueller, 1881)	F[I]	potimirim
Syncaris pacifica (Holmes, 1895)[2]	F	California freshwater shrimp
Syncaris pasadenae (Kingsley, 1896)	F	Pasadena shrimp

[2]Endangered species. California Administrative Code, Title 14, Section 670.5 (1988); U.S. Federal Register 53(210): 43884-43889 (31 Oct. 1988).

SCIENTIFIC NAME	OCCURRENCE	COMMON NAME

Oplophoridae—deepsea shrimps

Acanthephyra acanthitelsonis Bate, 1888	A	
Acanthephyra acutifrons Bate, 1888	A	
Acanthephyra armata A. Milne Edwards, 1881	A	
Acanthephyra brevirostris Smith, 1885	A	
Acanthephyra chacei Krygier and Forss, 1981	P	
Acanthephyra curtirostris Wood-Mason, 1891	P	peaked shrimp
Acanthephyra eximia Smith, 1884	A	
Acanthephyra pelagica (Risso, 1816)	A	
Acanthephyra purpurea A. Milne Edwards, 1881	A	
Acanthephyra stylorostratis (Bate, 1888)	A	
Ephyrina benedicti Smith, 1885	A	
Ephyrina bifida Stephensen, 1923	A	
Ephyrina figueirai Crosnier and Forest, 1973	A	
Ephyrina ombango Crosnier and Forest, 1973	A	
Hymenodora acanthitelsonis Wasmer, 1972	P	
Hymenodora frontalis Rathbun, 1902	P	Pacific ambereye
Hymenodora glacialis (Buchholz, 1874)	A–P	northern ambereye
Hymenodora gracilis Smith, 1886	A–P	
Meningodora compsa (Chace, 1940)	A	
Meningodora marptocheles (Chace, 1940)	A	
Meningodora mollis Smith, 1882	A	
Meningodora vesca (Smith, 1886)	A	
Notostomus distirus Chace, 1940	A	
Notostomus elegans A. Milne Edwards, 1881	A	
Notostomus gibbosus A. Milne Edwards, 1881	A	
Notostomus japonicus Bate, 1888	P	Japanese spinyridge
Notostomus robustus Smith, 1884	A	
Oplophorus gracilirostris A. Milne Edwards, 1881	A	
Oplophorus grimaldii Coutière, 1905	A	
Oplophorus spinicauda A. Milne Edwards, 1883	A	
Oplophorus spinosus (Brullé, 1839)	A	
Systellaspis affinis (Faxon, 1896)	A	
Systellaspis braueri (Balss, 1914)	A–P	Quayle spinytail
Systellaspis cristata (Faxon, 1893)	P	Krygier spinytail
Systellaspis debilis (A. Milne Edwards, 1881)	A	
Systellaspis pellucida (Filhol, 1885)	A	

Nematocarcinidae

Nematocarcinus cursor A. Milne Edwards, 1881	A	
Nematocarcinus ensifer (Smith, 1882)	A	
Nematocarcinus rotundus Crosnier and Forest, 1973	A	

Superfamily Stylodactyloida

Stylodactylidae

Stylodactylus licinus Chace, 1983	A	

SCIENTIFIC NAME	OCCURRENCE	COMMON NAME

Superfamily Pasiphaeoidea

Pasiphaeidae—glass shrimps

Leptochela bermudensis Gurney, 1939	A	.... Bermuda comb shrimp
Leptochela carinata Ortmann, 1893	A	.... carinate glass shrimp
Leptochela papulata Chace, 1976	A	.... light glass shrimp
Leptochela serratorbita Bate, 1888	A	.... combclaw shrimp
Parapasiphae compta Smith, 1884	A	....
Parapasiphae cristata Smith, 1884	A	....
Parapasiphae macrodactyla Chace, 1939	A	....
Parapasiphae serrata Rathbun, 1902	P	....
Parapasiphae sulcatifrons Smith, 1884	A–P	.... grooveback shrimp
Pasiphaea affinis Rathbun, 1902	P	....
Pasiphaea chacei Yaldwyn, 1962	P	....
Pasiphaea corteziana Rathbun, 1902	P	....
Pasiphaea emarginata Rathbun, 1902	P	....
Pasiphaea magna Faxon, 1893	P	....
Pasiphaea merriami Schmitt, 1931	A	.... ghost comb shrimp
Pasiphaea multidentata Esmark, 1866	A–P	.... pink glass shrimp
Pasiphaea pacifica Rathbun, 1902	P	.... Pacific glass shrimp
Pasiphaea sivado (Risso, 1816)	A	....
Pasiphaea tarda Krøyer, 1845	A–P	.... crimson pasiphaeid

Superfamily Rhynchocinetoidea

Bresiliidae

Discias atlanticus Gurney, 1939	A	....
Discias serratirostris Lebour, 1949	A	....
Discias vernbergi Boothe and Heard, 1987	A	....
Pseudocheles chacei Kensley, 1983	A	....

Eugonatonotidae

Eugonatonotus crassus (A. Milne Edwards, 1881) ...	A	....

Rhynchocinetidae

Rhynchocinetes rigens Gordon, 1936	A	.... mechanical shrimp

Superfamily Palaemonoidea

Campylonotidae

Bathypalaemonella serratipalma Pequegnat, 1970	A	....
Bathypalaemonella texana Pequegnat, 1970	A	....

Gnathophyllidae—bumblebee shrimps

Gnathophylloides mineri Schmitt, 1933	A	.... squat urchin shrimp
Gnathophyllum americanum Guérin-Méneville, 1855 ..	A	.... bumblebee shrimp
Gnathophyllum circellus Manning, 1963	A	.... circled shrimp
Gnathophyllum modestum Hay, 1917	A	...... spotted bumblebee shrimp

SCIENTIFIC NAME	OCCURRENCE	COMMON NAME
Palaemonidae		
Anchistioides antiguensis (Schmitt, 1924)	A	....
Brachycarpus biunguiculatus (Lucas, 1849)	A–P	.... twoclaw shrimp
Leander paulensis Ortmann, 1897	A	....
Leander tenuicornis (Say, 1818)	A	.... brown grass shrimp
Lipkebe holthuisi Chace, 1969	A	....
Macrobrachium acanthurus (Wiegmann, 1836)	A–F	.... cinnamon river shrimp
Macrobrachium carcinus (Linnaeus, 1758)	A–F	.... bigclaw river shrimp
Macrobrachium crenulatum Holthuis, 1950	A–F	....
Macrobrachium faustinum (de Saussure, 1857)	A–F	....
Macrobrachium heterochirus (Wiegmann, 1836)	A–F	.... cascade river prawn
Macrobrachium ohione (Smith, 1874)	A–F	.... Ohio shrimp
Macrobrachium olfersi (Wiegmann, 1836)	A–F	.... bristled river shrimp
Neopontonides beaufortensis (Borradaile, 1920)	A	.... seawhip shrimp
Neopontonides chacei Heard, 1986	A	....
Palaemon floridanus Chace, 1942	A	.... Florida grass shrimp
Palaemon macrodactylus Rathbun, 1902	P	.... oriental shrimp
Palaemon northropi (Rankin, 1898)	A	....
Palaemon ritteri Holmes, 1895	P	.... barred grass shrimp
Palaemonella holmesi (Nobili, 1907)	P	.... southern algae shrimp
Palaemonetes antrorum Benedict, 1896	F	.... Balcones cave shrimp
Palaemonetes cummingi Chace, 1954	F	.. Squirrel Chimney Cave shrimp
Palaemonetes hiltoni Schmitt, 1921	P	.... Hilton shrimp
Palaemonetes holthuisi Strenth, 1976	F	.... Purgatory cave shrimp
Palaemonetes intermedius Holthuis, 1949	A	.... brackish grass shrimp
Palaemonetes kadiakensis Rathbun, 1902	F	.... Mississippi grass shrimp
Palaemonetes paludosus (Gibbes, 1850)	F	.... riverine grass shrimp
Palaemonetes pugio Holthuis, 1949	A	.... daggerblade grass shrimp
Palaemonetes texanus Strenth, 1976	F	.... Texas river shrimp
Palaemonetes vulgaris (Say, 1818)	A	.... marsh grass shrimp
Periclimenaeus ascidiarum Holthuis, 1951	A	....
Periclimenaeus atlanticus (Rathbun, 1901)	A	....
Periclimenaeus bermudensis (Armstrong, 1940)	A	....
Periclimenaeus bredini Chace, 1972	A	....
Periclimenaeus caraibicus Holthuis, 1971	A	....
Periclimenaeus chacei Abele, 1971	A	....
Periclimenaeus maxillulidens (Schmitt, 1936)	A	....
Periclimenaeus pearsei (Schmitt, 1932)	A	....
Periclimenaeus perlatus (Boone, 1930)	A	....
Periclimenaeus schmitti Holthuis, 1951	A	.... Tortugas bigclaw
Periclimenaeus wilsoni (Hay, 1917)	A	.... clear sponge shrimp
Periclimenes americanus (Kingsley, 1878)	A	.... American grass shrimp
Periclimenes harringtoni Lebour, 1949	A	.... Bermuda shrimp
Periclimenes infraspinis (Rathbun, 1902)	P	....
Periclimenes iridescens Lebour, 1949	A	....
Periclimenes longicaudatus (Stimpson, 1860)	A	.... longtail grass shrimp
Periclimenes magnus Holthuis, 1951	A	....
Periclimenes pandionis Holthuis, 1951	A	....
Periclimenes pedersoni Chace, 1958	A	.... Pederson cleaner shrimp
Periclimenes perryae Chace, 1942	A	....
Periclimenes rathbunae Schmitt, 1924	A	....

SCIENTIFIC NAME	OCCURRENCE	COMMON NAME
Periclimenes tenellus (Smith, 1882)	A	....
Periclimenes yucatanicus (Ives, 1891)	A	.... spotted cleaner shrimp
Pontonia californiensis Rathbun, 1902	P	.... sea squirt shrimp
Pontonia domestica Gibbes, 1850	A	.... Atlantic pen shrimp
Pontonia margarita Smith, 1869	A	.... pearl oyster shrimp
Pontonia mexicana Guérin-Méneville, 1856	A	.... Caribbean pen shrimp
Pontonia unidens Kingsley, 1880	A	....
Pontoniopsis paulae Gore, 1981	A	....
Pseudocoutierea antillensis Chace, 1972	A	....
Pseudocoutierea elegans Holthuis, 1951	P	....
Tuleariocaris neglecta Chace, 1969	A	.... black urchin shrimp
Typton carneus Holthuis, 1951	A	....
Typton distinctus Chace, 1972	A	....
Typton gnathophylloides Holthuis, 1951	A	....
Typton prionurus Holthuis, 1951	A	....
Typton tortugae McClendon, 1911	A	.... pale sponge shrimp
Typton vulcanus Holthuis, 1951	A	....
Veleroniopsis kimallynae Gore, 1981	A	....

Superfamily Psalidopodoidea

Psalidopodidae—scissorfoot shrimps

Psalidopus barbouri Chace, 1939	A	....

Superfamily Alpheoidea

Alpheidae—snapping shrimps

Alpheopsis equidactylus (Lockington, 1877)	P	....
Alpheopsis harperi Wicksten, 1984	A	....
Alpheopsis labis Chace, 1972	A	..... tongclaw snapping shrimp
Alpheopsis trispinosa (Stimpson, 1860)	A	....
Alpheus amblyonyx Chace, 1972	A	..... bluntclaw snapping shrimp
Alpheus armatus Rathbun, 1901	A	....
Alpheus armillatus H. Milne Edwards, 1837	A	.... banded snapping shrimp
Alpheus bahamensis Rankin, 1898	A	....
Alpheus barbara Lockington, 1878	P	....
Alpheus beanii Verrill, 1922	A	....
Alpheus belli Coutière, 1898	A	....
Alpheus bellimanus Lockington, 1877	P	.... olive snapping shrimp
Alpheus bouvieri A. Milne Edwards, 1878	A	....
Alpheus californiensis Holmes, 1900	P	.... mudflat snapping shrimp
Alpheus candei Guérin-Méneville, 1855	A	....
Alpheus clamator Lockington, 1877	P	.... twistclaw pistol shrimp
Alpheus cristulifrons Rathbun, 1900	A	.... dotted snapping shrimp
Alpheus cylindricus Kingsley, 1878	A	.... cylindrical snapping shrimp
Alpheus estuariensis Christoffersen, 1984	A	....
Alpheus floridanus Kingsley, 1878	A	.... sand snapping shrimp
Alpheus formosus Gibbes, 1850	A	.... striped snapping shrimp
Alpheus heterochaelis Say, 1818	A	.... bigclaw snapping shrimp
Alpheus normanni Kingsley, 1878	A	.... green snapping shrimp
Alpheus nuttingi (Schmitt, 1924)	A	....

SCIENTIFIC NAME	OCCURRENCE	COMMON NAME
Alpheus paracrinitus Miers, 1881	A	.... smoothclaw snapping shrimp
Alpheus peasei (Armstrong, 1940)	A	..... orangetail snapping shrimp
Alpheus schmitti Chace, 1972	A	
Alpheus thomasi Hendrix and Gore, 1973	A	
Alpheus viridari (Armstrong, 1949)	A	
Alpheus websteri Kingsley, 1880	A	
Automate dolichognatha de Man, 1888	A	
Automate evermanni Rathbun, 1901	A	
Automate gardineri Coutière, 1902	A	
Automate rectifrons Chace, 1972	A	
Betaeus ensenadensis Glassell, 1938	P	.... mudflat visored shrimp
Betaeus gracilis Hart, 1964	P	.... kelp visored shrimp
Betaeus harfordi (Kingsley, 1878)	P	.... abalone visored shrimp
Betaeus harrimani Rathbun, 1904	P	.... northern hooded shrimp
Betaeus longidactylus Lockington, 1877	P	.... visored shrimp
Betaeus macginitieae Hart, 1964	P	.... urchin visored shrimp
Betaeus setosus Hart, 1964	P	.... fuzzy hooded shrimp
Fenneralpheus chacei Felder and Manning, 1986	A	
Leptalpheus forceps Williams, 1965	A	
Metalpheus rostratipes (Pocock, 1890)	A	
Salmoneus cavicola Felder and Manning, 1986	A	
Synalpheus agelas Pequegnat and Heard, 1979	A	
Synalpheus apioceros Coutière, 1909	A	
Synalpheus bousfieldi Chace, 1972	A	
Synalpheus brevicarpus (Herrick, 1891)	A	
Synalpheus brooksi Coutière, 1909	A	
Synalpheus disparodigitus Armstrong, 1949	A	
Synalpheus fritzmuelleri Coutière, 1909	A	...... speckled snapping shrimp
Synalpheus goodei Coutière, 1909	A	
Synalpheus heardi Dardeau, 1984	A	
Synalpheus hemphilli Coutière, 1909	A	
Synalpheus herricki Coutière, 1909	A	
Synalpheus lockingtoni Coutière, 1909	P	.... littoral pistol shrimp
Synalpheus longicarpus (Herrick, 1891)	A	
Synalpheus macclendoni Coutière, 1910	A	
Synalpheus minus (Say, 1818)	A	.... minor snapping shrimp
Synalpheus pandionis Coutière, 1909	A	.... turtlegrass snapping shrimp
Synalpheus paraneptunus Coutière, 1909	A	
Synalpheus pectiniger Coutière, 1907	A	..... loggerhead snapping shrimp
Synalpheus rathbunae Coutière, 1909	A	
Synalpheus sanctithomae Coutière, 1909	A	
Synalpheus scaphoceris Coutière, 1910	A	
Synalpheus townsendi Coutière, 1909	A	.... Townsend snapping shrimp
Thunor simus (Guérin-Méneville, 1855)	A	

Hippolytidae

Bythocaris gracilis Smith, 1885	A	
Bythocaris leucopis G. O. Sars, 1879	A	
Bythocaris nana Smith, 1885	A	
Bythocaris payeri (Heller, 1875)	A	
Bythocaris simplicirostris G. O. Sars, 1869	A	

SCIENTIFIC NAME	OCCURRENCE	COMMON NAME
Caridion gordoni (Bate, 1858)	A	bigclaw hippolyte
Eualus avinus (Rathbun, 1899)	P	beaked eualid
Eualus barbatus (Rathbun, 1899)	P	barbed eualid
Eualus berkeleyorum Butler, 1971	P	Berkeley eualid
Eualus biunguis (Rathbun, 1902)	P	deepsea eualid
Eualus fabricii (Krøyer, 1841)	A–P	Arctic eualid
Eualus gaimardii (H. Milne Edwards, 1837)	A–P	circumpolar eualid
Eualus herdmani (Walker, 1898)	P	Herdman eualid
Eualus lineatus Wicksten and Butler, 1983	P	striped eualid
Eualus macilentus (Krøyer, 1841)	A–P	Greenland shrimp
Eualus macropthalmus (Rathbun, 1902)	P	bigeye eualid
Eualus pusiolus (Krøyer, 1841)	A–P	doll eualid
Eualus stoneyi (Rathbun, 1902)	A–P	comb-beak eualid
Eualus suckleyi (Stimpson, 1864)	P	shortscale eualid
Eualus townsendi (Rathbun, 1902)	P	Townsend eualid
Exhippolysmata oplophoroides (Holthuis, 1948)	A	redleg humpback shrimp
Heptacarpus brachydactylus (Rathbun, 1902)	P	island coastal shrimp
Heptacarpus brevirostris (Dana, 1852)	P	stout coastal shrimp
Heptacarpus camtschaticus (Stimpson, 1860)	P	northern coastal shrimp
Heptacarpus carinatus Holmes, 1900	P	smalleye coastal shrimp
Heptacarpus decorus (Rathbun, 1902)	P	elegant coastal shrimp
Heptacarpus flexus (Rathbun, 1902)	P	slenderbeak coastal shrimp
Heptacarpus franciscanus (Schmitt, 1921)	P	Franciscan coastal shrimp
Heptacarpus kincaidi (Rathbun, 1902)	P	Kincaid coastal shrimp
Heptacarpus littoralis Butler, 1980	P	bigeye coastal shrimp
Heptacarpus maxillipes (Rathbun, 1902)	P	Aleutian coastal shrimp
Heptacarpus moseri (Rathbun, 1902)	P	Alaska coastal shrimp
Heptacarpus palpator (Owen, 1839)	P	intertidal coastal shrimp
Heptacarpus paludicola Holmes, 1900	P	California coastal shrimp
Heptacarpus pictus (Stimpson, 1871)	P	redbanded clear shrimp
Heptacarpus pugettensis Jensen, 1983	P	Puget coastal shrimp
Heptacarpus sitchensis (Brandt, 1851)	P	Sitka coastal shrimp
Heptacarpus stimpsoni Holthuis, 1947	P	Stimpson coastal shrimp
Heptacarpus stylus (Stimpson, 1864)	P	stiletto coastal shrimp
Heptacarpus taylori (Stimpson, 1857)	P	Taylor coastal shrimp
Heptacarpus tenuissimus Holmes, 1900	P	slender coastal shrimp
Heptacarpus tridens (Rathbun, 1902)	P	threespine coastal shrimp
Hippolyte californiensis Holmes, 1895	P	California green shrimp
Hippolyte clarki Chace, 1951	P	kelp humpback shrimp
Hippolyte coerulescens (Fabricius, 1775)	A	cerulean sargassum shrimp
Hippolyte curacaoensis Schmitt, 1924	A	
Hippolyte nicholsoni Chace, 1952	A	
Hippolyte pleuracantha (Stimpson, 1871)	A	false zostera shrimp
Hippolyte zostericola (Smith, 1873)	A	zostera shrimp
Latreutes fucorum (Fabricius, 1798)	A	slender sargassum shrimp
Latreutes inermis Chace, 1972	A	
Latreutes parvulus (Stimpson, 1866)	A	sargassum shrimp
Lebbeus brandti (Brazhnikov, 1907)	P	
Lebbeus catalepsis Jensen, 1987	P	cataleptic lebbeid
Lebbeus grandimana (Brazhnikov, 1907)	P	candy-striped shrimp
Lebbeus groenlandicus (Fabricius, 1775)	A–P	spiny lebbeid

SCIENTIFIC NAME	OCCURRENCE	COMMON NAME
Lebbeus lagunae (Schmitt, 1921)	P	Laguna lebbeid
Lebbeus microceros (Krøyer, 1841	A	
Lebbeus polaris (Sabine, 1821)	A–P	polar lebbeid
Lebbeus possjeticus Kobyakova, 1967	P	Possjet lebbeid
Lebbeus schrencki (Brazhnikov, 1907)	P	Okhotsk lebbeid
Lebbeus unalaskensis (Rathbun, 1902)	P	
Lebbeus vicinus (Rathbun, 1902)	P	offshore lebbeid
Lebbeus washingtonianus (Rathbun, 1902)	P	slope lebbeid
Lebbeus zebra (Leim, 1921)	A–P	zebra lebbeid
Lysmata californica (Stimpson, 1866)	P	red rock shrimp
Lysmata grabhami (Gordon, 1935)	A	cleaner shrimp
Lysmata intermedia (Kingsley, 1878)	A	
Lysmata rathbunae Chace, 1970	A	
Lysmata wurdemanni (Gibbes, 1850)	A	peppermint shrimp
Merhippolyte americana Holthuis, 1961	A	
Spirontocaris affinis (Owen, 1866)	P	
Spirontocaris arcuata Rathbun, 1902	P	Rathbun blade shrimp
Spirontocaris dalli Rathbun, 1902	P	Dall blade shrimp
Spirontocaris holmesi Holthuis, 1947	P	slender blade shrimp
Spirontocaris lamellicornis (Dana, 1852)	P	
Spirontocaris liljeborgii (Danielssen, 1859)	A–P	friendly blade shrimp
Spirontocaris murdochi Rathbun, 1902	P	
Spirontocaris ochotensis (Brandt, 1851)	P	oval blade shrimp
Spirontocaris phippsii (Krøyer, 1841)	A–P	punctate blade shrimp
Spirontocaris prionota (Stimpson, 1864)	P	deep blade shrimp
Spirontocaris sica Rathbun, 1902	P	offshore blade shrimp
Spirontocaris snyderi Rathbun, 1902	P	Snyder blade shrimp
Spirontocaris spinus (Sowerby, 1805)	A–P	parrot shrimp
Spirontocaris truncata Rathbun, 1902	P	blunt blade shrimp
Thor amboinensis (de Man, 1888)	A–P	squat anemone shrimp
Thor dobkini Chace, 1972	A	squat grass shrimp
Thor floridanus Kingsley, 1878	A	bryozoan shrimp
Thor manningi Chace, 1972	A	Manning grass shrimp
Tozeuma carolinense Kingsley, 1878	A	arrow shrimp
Tozeuma cornutum A. Milne Edwards, 1881	A	
Tozeuma serratum A. Milne Edwards, 1881	A	
Trachycaris restricta (A. Milne Edwards, 1878)	A	

Ogyrididae—longeye shrimps

Ogyrides alphaerostris (Kingsley, 1880)	A	estuarine longeye shrimp
Ogyrides hayi Williams, 1981	A	sand longeye shrimp

Processidae—night shrimps

Ambidexter panamensis Abele, 1972	P	
Ambidexter symmetricus Manning and Chace, 1971	A	
Nikoides schmitti Manning and Chace, 1971	A	
Processa bermudensis (Rankin, 1900)	A	Bermuda night shrimp
Processa fimbriata Manning and Chace, 1971	A	grass night shrimp
Processa guyanae Holthuis, 1959	A	
Processa hemphilli Manning and Chace, 1971	A	

SCIENTIFIC NAME	OCCURRENCE	COMMON NAME
Processa profunda Manning and Chace, 1971	A	....
Processa riveroi Manning and Chace, 1971	A	....
Processa vicina Manning and Chace, 1971	A	....

Superfamily Pandaloidea

Pandalidae

Dichelopandalus leptocerus (Smith, 1881)	A	.... bristled longbeak
Heterocarpus ensifer A. Milne Edwards, 1881	A	.... armed nylon shrimp
Heterocarpus oryx A. Milne Edwards, 1881	A	....
Pandalopsis aleutica Rathbun, 1902	P	.... Aleutian bigeye
Pandalopsis ampla Bate, 1888	P	.... deepwater bigeye
Pandalopsis dispar Rathbun, 1902	P	.... sidestriped shrimp
Pandalopsis longirostris Rathbun, 1902	P	.... northern longbeak
Pandalus borealis Krøyer, 1838	A–P	.... northern shrimp
Pandalus danae Stimpson, 1857	P	.... dock shrimp
Pandalus goniurus Stimpson, 1860	P	.... humpy shrimp
Pandalus gurneyi Stimpson, 1871	P	.... California longbeak
Pandalus hypsinotus Brandt, 1851	P	.... coonstriped shrimp
Pandalus jordani Rathbun, 1902	P	.... ocean shrimp
Pandalus montagui Leach, 1814	A	.... Aesop shrimp
Pandalus platyceros Brandt, 1851	P	.... spot shrimp
Pandalus propinquus G. O. Sars, 1869	A	....
Pandalus stenolepis Rathbun, 1902	P	.... roughpatch shrimp
Pandalus tridens Rathbun, 1902	P	.... yellowleg pandalid
Pantomus parvulus A. Milne Edwards, 1883	A	.... hinged longbeak
Plesionika acanthonotus (Smith, 1882)	A	.... striped shrimp
Plesionika edwardsii (Brandt, 1851)	A	.... soldier striped shrimp
Plesionika ensis (A. Milne Edwards, 1881)	A	.... gladiator striped shrimp
Plesionika escatilis (Stimpson, 1860)	A	....
Plesionika holthuisi Crosnier and Forest, 1968	A	....
Plesionika longicauda (Rathbun, 1902)	A	....
Plesionika longipes (A. Milne Edwards, 1881)	A	....
Plesionika martia (A. Milne Edwards, 1883)	A	.... golden shrimp
Plesionika mexicana Chace, 1937	P	.... Mexican longbeak
Plesionika miles (A. Milne Edwards, 1883)	A	....
Plesionika polyacanthomerus Pequegnat, 1970	A	....
Plesionika sanctaecatalinae Wicksten, 1983	P	....
Plesionika tenuipes (Smith, 1881)	A	....
Plesionika williamsi Forest, 1964	A	....
Plesionika willisi (Pequegnat, 1970)	A	....
Stylopandalus richardi (Coutière, 1905)	A	....

Superfamily Crangonoidea

Crangonidae

Argis alaskensis (Kingsley, 1882)	P	.... Alaska argid
Argis californiensis (Rathbun, 1902)	P	.... slope argid
Argis crassa (Rathbun, 1899)	P	.... rough argid
Argis dentata (Rathbun, 1902)	A–P	.... Arctic argid
Argis lar (Owen, 1839)	P	.... kuro shrimp

SCIENTIFIC NAME	OCCURRENCE	COMMON NAME
Argis levior (Rathbun, 1902)	P	Nelson argid
Argis ovifer (Rathbun, 1902)	P	spliteye argid
Crangon alba Holmes, 1900	P	stout crangon
Crangon communis Rathbun, 1899	P	twospine crangon
Crangon dalli Rathbun, 1902	P	ridged crangon
Crangon franciscorum Stimpson, 1856	P	California bay shrimp
Crangon handi Kuris and Carlton, 1977	P	
Crangon holmesi Rathbun, 1902	P	Holmes bay shrimp
Crangon nigricauda Stimpson, 1856	P	blacktail bay shrimp
Crangon nigromaculata Lockington, 1877	P	blackspotted bay shrimp
Crangon septemspinosa Say, 1818	A	sevenspine bay shrimp
Lissocrangon stylirostris (Holmes, 1900)	P	smooth bay shrimp
Mesocrangon intermedia (Stimpson, 1860)	P	northern spinyhead
Mesocrangon munitella (Walker, 1898)	P	miniature spinyhead
Metracrangon acclivis (Rathbun, 1902)	P	forked spinyhead
Metacrangon jacqueti (A. Milne Edwards, 1881)	A	
Metacrangon munita (Dana, 1852)	P	coastal spinyhead
Metacrangon spinirostris (Rathbun, 1902)	P	
Metacrangon spinosissima (Rathbun, 1902)	P	southern spinyhead
Metacrangon variabilis (Rathbun, 1902)	P	deepsea spinyhead
Neocrangon abyssorum Rathbun, 1902	P	abyssal crangon
Neocrangon alaskensis Lockington, 1877	P	Alaska bay shrimp
Neocrangon communis (Rathbun, 1899)	P	gray shrimp
Neocrangon resima (Rathbun, 1902)	P	
Neocrangon zacae (Chace, 1937)	P	
Paracrangon echinata Dana, 1852	P	horned shrimp
Parapontocaris caribbaea (Boone, 1927)	A	
Parapontocaris vicina (Dardeau and Heard, 1983)	A	
Philocheras gorei (Dardeau, 1980)	A	
Pontophilus abyssi Smith, 1884	A	
Pontophilus brevirostris Smith, 1881	A	
Pontophilus gorei Dardeau, 1980	A	
Pontophilus gracilis Smith, 1882	A	
Pontophilus norvegicus (M. Sars, 1861)	A	Norwegian shrimp
Rhynocrangon alata (Rathbun, 1902)	P	saddleback shrimp
Rhynocrangon sharpi (Ortmann, 1895)	P	
Sabinea hystrix (A. Milne Edwards, 1881)	A	
Sabinea sarsii Smith, 1879	A	Sars shrimp
Sabinea septemcarinata (Sabine, 1824)	A–P	sevenline shrimp
Sabinea tridentata Pequegnat, 1970	A	
Sclerocrangon boreas (Phipps, 1774)	A–P	sculptured shrimp
Sclerocrangon ferox (G. O. Sars, 1877)	A	

Glyphocrangonidae—armored shrimps

Glyphocrangon aculeata A. Milne Edwards, 1881	A	
Glyphocrangon alispina Chace, 1939	A	
Glyphocrangon haematonotus Holthuis, 1971	A	
Glyphocrangon longirostris (Smith, 1882)	A	
Glyphocrangon longleyi Schmitt, 1931	A	
Glyphocrangon nobilis A. Milne Edwards, 1881	A	
Glyphocrangon sculpta (Smith, 1882)	A	

SCIENTIFIC NAME	OCCURRENCE	COMMON NAME
Glyphocrangon spinicauda A. Milne Edwards, 1881 ..	A	....
Glyphocrangon spinulosa Faxon, 1893	P	....
Glyphocrangon vicaria Faxon, 1896	P	....

INFRAORDER ASTACIDEA

Superfamily Nephropoidea

Nephropidae—clawed lobsters

Acanthacaris caeca (A. Milne Edwards, 1881)	A	.... Atlantic deepsea lobster
Eunephrops manningi Holthuis, 1974	A	....
Homarus americanus H. Milne Edwards, 1837	A	.... American lobster
Metanephrops binghami (Boone, 1927)	A	.... Caribbean lobsterette
Nephropsis aculeata Smith, 1881	A	.... Florida lobsterette
Nephropsis agassizii A. Milne Edwards, 1880	A	.... prickly lobsterette
Nephropsis rosea Bate, 1888	A	.... rosy lobsterette
Nephropsis neglecta Holthuis, 1974	A	....

Superfamily Astacoidea

Astacidae—crayfishes

Pacifastacus connectens (Faxon, 1914)	F	....
Pacifastacus fortis (Faxon, 1914)[3]	F	.... Shasta crayfish
Pacifastacus gambelii (Girard, 1852)	F	....
Pacifastacus leniusculus (Dana, 1852)	F	.... signal crayfish
Pacifastacus nigrescens (Stimpson, 1857)	F	.... sooty crayfish

Cambaridae—crayfishes

Barbicambarus cornutus (Faxon, 1884)	F	.... bottlebrush crayfish
Bouchardina robisoni Hobbs, 1977	F	....
Cambarellus blacki Hobbs, 1980	F	.... cypress crayfish
Cambarellus diminutus Hobbs, 1945	F	.... least crayfish
Cambarellus lesliei Fitzpatrick and Laning, 1976	F	....
Cambarellus ninae Hobbs, 1950	F	....
Cambarellus puer Hobbs, 1941	F	....
Cambarellus schmitti Hobbs, 1942	F	....
Cambarellus shufeldtii (Faxon, 1884)	F	.... Cajun dwarf crayfish
Cambarellus texanus Albaugh and Black, 1973	F	....
Cambarus acanthura Hobbs, 1981	F	....
Cambarus acuminatus Faxon, 1884	F	....
Cambarus asperimanus Faxon, 1914	F	....
Cambarus bartonii (Fabricius, 1798)	F	..... Appalachian brook crayfish
Cambarus batchi Schuster, 1976	F	.... bluegrass crayfish
Cambarus bouchardi Hobbs, 1970	F	.... Big South Fork crayfish
Cambarus brachydactylus Hobbs, 1953	F	....
Cambarus buntingi Bouchard, 1973	F	....
Cambarus carolinus (Erichson, 1846)	F	....

---

[3]Endangered species. California Administrative Code, Title 4, Section 670.5 (1988); U.S. Federal Register 53(190): 38460-38465 (30 Sept. 1988).

SCIENTIFIC NAME	OCCURRENCE	COMMON NAME
Cambarus catagius Hobbs and Perkins, 1967	F	.. Greensboro burrowing crayfish
Cambarus causeyi Reimer, 1966	F	....
Cambarus chasmodactylus James, 1966	F	.... New River crayfish
Cambarus chaugaensis Prins and Hobbs, 1969	F	....
Cambarus conasaugaensis Hobbs and Hobbs, 1962 ..	F	....
Cambarus coosae Hobbs, 1981	F	....
Cambarus coosawattae Hobbs, 1981	F	....
Cambarus cracens Bouchard and Hobbs, 1976	F	....
Cambarus crinipes Bouchard, 1973	F	....
Cambarus cryptodytes Hobbs, 1941	F	... Dougherty Plain cave crayfish
Cambarus cumberlandensis Hobbs and Bouchard, 1973	F	....
Cambarus cymatilis Hobbs, 1970	F	....
Cambarus deweesae Bouchard and Etnier, 1979	F	.... valley flame crayfish
Cambarus diogenes Girard, 1852	F	.... devil crawfish
Cambarus distans Rhoades, 1944	F	....
Cambarus dubius Faxon, 1884	F	....
Cambarus englishi Hobbs and Hall, 1972	F	....
Cambarus extraneus Hagen, 1870	F	.... Chickamauga crayfish
Cambarus fasciatus Hobbs, 1981	F	....
Cambarus friaufi Hobbs, 1953	F	....
Cambarus gentryi Hobbs, 1970	F	....
Cambarus georgiae Hobbs, 1981	F	...... Little Tennessee crayfish
Cambarus girardianus Faxon, 1884	F	....
Cambarus graysoni Faxon, 1914	F	....
Cambarus halli Hobbs, 1968	F	....
Cambarus hamulatus (Cope, 1881)	F	....
Cambarus harti Hobbs, 1981	F	.... piedmont blue burrower
Cambarus hiwasseensis Hobbs, 1981	F	....
Cambarus howardi Hobbs and Hall, 1969	F	....
Cambarus hubbsi Creaser, 1931	F	....
Cambarus hubrichti Hobbs, 1951	F	....
Cambarus jonesi Hobbs and Barr, 1960	F	.... Alabama cave crayfish
Cambarus laevis Faxon, 1914	F	....
Cambarus latimanus (Le Conte, 1856)	F	....
Cambarus longirostris Faxon, 1885	F	....
Cambarus longulus Girard, 1852	F	....
Cambarus manningi Hobbs, 1981	F	....
Cambarus miltus Fitzpatrick, 1978	F	.... rusty grave digger
Cambarus monongalensis Ortmann, 1905	F	....
Cambarus nerterius Hobbs, 1964	F	....
Cambarus nodosus Bouchard and Hobbs, 1976	F	....
Cambarus obeyensis Hobbs and Shoup, 1947	F	.... Obey crayfish
Cambarus obstipus Hall, 1959	F	....
Cambarus ornatus Rhoades, 1944	F	....
Cambarus ortmanni Williamson, 1907	F	....
Cambarus parrishi Hobbs, 1981	F	....
Cambarus parvoculus Hobbs and Shoup, 1947	F	....
Cambarus pristinus Hobbs, 1965	F	....
Cambarus pyronotus Bouchard, 1978	F	.... fireback crayfish
Cambarus reburrus Prins, 1968	F	.... French Broad crayfish

SCIENTIFIC NAME	OCCURRENCE	COMMON NAME
Cambarus reduncus Hobbs, 1956	F	
Cambarus reflexus Hobbs, 1981	F	
Cambarus robustus Girard, 1852	F	
Cambarus rusticiformis Rhoades, 1944	F	
Cambarus sciotensis Rhoades, 1944	F	
Cambarus scotti Hobbs, 1981	F	
Cambarus setosus Faxon, 1889	F	
Cambarus speciosus Hobbs, 1981	F	
Cambarus sphenoides Hobbs, 1968	F	
Cambarus spicatus Hobbs, 1956	F	
Cambarus striatus Hay, 1902	F	
Cambarus strigosus Hobbs, 1981	F	
Cambarus tartarus Hobbs and Cooper, 1972	F	Oklahoma cave crayfish
Cambarus tenebrosus Hay, 1902	F	
Cambarus truncatus Hobbs, 1981	F	Oconee burrowing crayfish
Cambarus unestami Hobbs and Hall, 1969	F	
Cambarus veteranus Faxon, 1914	F	
Cambarus zophonastes Hobbs and Bedinger, 1964	F	Hell Creek cave crayfish
Distocambarus carlsoni Hobbs, 1983	F	mimic crayfish
Distocambarus crockeri Hobbs and Carlson, 1983	F	
Distocambarus devexus (Hobbs, 1981)	F	
Distocambarus yongineri Hobbs and Carlson, 1985	F	
Fallicambarus byersi (Hobbs, 1941)	F	lavender burrowing crayfish
Fallicambarus caesius Hobbs, 1975	F	
Fallicambarus danielae Hobbs, 1975	F	speckled burrowing crayfish
Fallicambarus devastator Hobbs and Whiteman, 1987	F	Texas prairie crayfish
Fallicambarus dissitus (Penn, 1955)	F	
Fallicambarus fodiens (Cottle, 1863)	F	
Fallicambarus harpi Hobbs and Robison, 1985	F	
Fallicambarus hedgpethi (Hobbs, 1948)	F	
Fallicambarus hortoni Hobbs and Fitzpatrick, 1970	F	Hatchie burrowing crayfish
Fallicambarus jeanae Hobbs, 1973	F	
Fallicambarus macneesei (Black, 1967)	F	
Fallicambarus orkytes Penn and Marlow, 1959	F	
Fallicambarus strawni (Reimer, 1966)	F	
Fallicambarus uhleri (Faxon, 1884)	F	
Faxonella beyeri (Penn, 1950)	F	
Faxonella blairi Hayes and Reimer, 1977	F	
Faxonella clypeata (Hay, 1899)	F	ditch fencing crayfish
Faxonella creaseri Walls, 1968	F	
Hobbseus attenuatus Black, 1969	F	Pearl riverlet crayfish
Hobbseus cristatus (Hobbs, 1955)	F	
Hobbseus orconectoides Fitzpatrick and Payne, 1968	F	Oktibbeha riverlet crayfish
Hobbseus petilus Fitzpatrick, 1977	F	Tombigbee riverlet crayfish
Hobbseus prominens (Hobbs, 1966)	F	
Hobbseus valleculus (Fitzpatrick, 1967)	F	Choctaw riverlet crayfish
Orconectes acares Fitzpatrick, 1965	F	
Orconectes alabamensis (Faxon, 1884)	F	
Orconectes australis (Rhoades, 1941)	F	
Orconectes barrenensis Rhoades, 1944	F	
Orconectes bisectus Rhoades, 1944	F	Crittenden crayfish

SCIENTIFIC NAME	OCCURRENCE	COMMON NAME
Orconectes causeyi Jester, 1967	F	
Orconectes chickasawae Cooper and Hobbs, 1980	F	
Orconectes compressus (Faxon, 1884)	F	
Orconectes cooperi Cooper and Hobbs, 1980	F	
Orconectes deanae Reimer and Jester, 1975	F	Conchas crayfish
Orconectes difficilis (Faxon, 1898)	F	painted crayfish
Orconectes erichsonianus (Faxon, 1898)	F	
Orconectes etnieri Bouchard and Bouchard, 1976	F	
Orconectes eupunctus Williams, 1952	F	
Orconectes forceps (Faxon, 1884)	F	
Orconectes harrisonii (Faxon, 1884)	F	
Orconectes hobbsi Penn, 1950	F	
Orconectes holti Cooper and Hobbs, 1980	F	
Orconectes hylas (Faxon, 1890)	F	
Orconectes illinoiensis Brown, 1956	F	
Orconectes immunis (Hagen, 1870)	F	calico crayfish
Orconectes incomptus Hobbs and Barr, 1972	F	Tennessee cave crayfish
Orconectes indianensis (Hay, 1896)	F	
Orconectes inermis Cope, 1872	F	
Orconectes jeffersoni Rhoades, 1944	F	Louisville crayfish
Orconectes kentuckiensis Rhoades, 1944	F	
Orconectes lancifer (Hagen, 1870)	F	
Orconectes leptogonpodus Hobbs, 1948	F	
Orconectes limosus (Rafinesque, 1817)	F	spinycheek crayfish
Orconectes longidigitus (Faxon, 1898)	F	
Orconectes luteus (Creaser, 1933)	F	
Orconectes macrus Williams, 1952	F	
Orconectes marchandi Hobbs, 1948	F	
Orconectes medius (Faxon, 1884)	F	
Orconectes meeki (Faxon, 1898)	F	
Orconectes menae (Creaser, 1933)	F	
Orconectes mirus (Ortmann, 1931)	F	
Orconectes mississippiensis (Faxon, 1884)	F	
Orconectes nais (Faxon, 1885)	F	
Orconectes nana Williams, 1952	F	
Orconectes neglectus (Faxon, 1885)	F	
Orconectes obscurus (Hagen, 1870)	F	
Orconectes ozarkae Williams, 1952	F	
Orconectes palmeri (Faxon, 1884)	F	
Orconectes pellucidus (Tellkampf, 1844)	F	
Orconectes perfectus Walls, 1972	F	
Orconectes peruncus (Creaser, 1931)	F	
Orconectes placidus (Hagen, 1870)	F	
Orconectes propinquus (Girard, 1852)	F	northern clearwater crayfish
Orconectes punctimanus (Creaser, 1933)	F	
Orconectes putnami (Faxon, 1884)	F	
Orconectes quadruncus (Creaser, 1933)	F	
Orconectes rafinesquei Rhoades, 1944	F	
Orconectes rhoadesi Hobbs, 1949	F	
Orconectes rusticus (Girard, 1852)	F	rusty crayfish
Orconectes sanbornii (Faxon, 1884)	F	

SCIENTIFIC NAME	OCCURRENCE	COMMON NAME
Orconectes saxatilis Bouchard and Bouchard, 1976 ...	F	Kiamichi crayfish
Orconectes shoupi Hobbs, 1948	F	Nashville crayfish
Orconectes sloanii (Bundy, 1876)	F	
Orconectes spinosus (Bundy, 1877)	F	
Orconectes stannardi Page, 1985	F	
Orconectes tricuspis Rhoades, 1944	F	
Orconectes validus (Faxon, 1914)	F	
Orconectes virginiensis Hobbs, 1951	F	Chowanoke crayfish
Orconectes virilis (Hagen, 1870)	F	virile crayfish
Orconectes williamsi Fitzpatrick, 1966	F	
Orconectes wrighti Hobbs, 1948	F	
Procambarus ablusus Penn, 1963	F	
Procambarus acherontis (Lönnberg, 1894)	F	Orlando cave crayfish
Procambarus acutissimus (Girard, 1852)	F	
Procambarus acutus (Girard, 1852)	F	white river crawfish
Procambarus advena (Le Conte, 1856)	F	
Procambarus alleni (Faxon, 1884)	F	
Procambarus ancylus Hobbs, 1972	F	
Procambarus angustatus (Le Conte, 1856)	F	
Procambarus apalachicolae Hobbs, 1942	F	
Procambarus barbatus (Faxon, 1890)	F	
Procambarus barbiger Fitzpatrick, 1978	F	Jackson prairie crayfish
Procambarus bivittatus Hobbs, 1942	F	ribbon crayfish
Procambarus blandingii (Harlan, 1830)	F	
Procambarus brazoriensis Albaugh, 1975	F	Brazoria crayfish
Procambarus capillatus Hobbs, 1971	F	
Procambarus caritus Hobbs, 1981	F	
Procambarus clarkii (Girard, 1852)	F	red swamp crawfish
Procambarus clemmeri Hobbs, 1975	F	
Procambarus cometes Fitzpatick, 1978	F	Mississippi flatwoods crayfish
Procambarus conus Fitzpatrick, 1978	F	Carrollton crayfish
Procambarus curdi Reimer, 1975	F	
Procambarus delicatus Hobbs and Franz, 1953	F	bigcheek cave crayfish
Procambarus dupratzii Penn, 1953	F	
Procambarus echinatus Hobbs, 1956	F	
Procambarus econfinae Hobbs, 1942	F	Panama City crayfish
Procambarus elegans Hobbs, 1969	F	
Procambarus enoplosternum Hobbs, 1947	F	
Procambarus epicyrtus Hobbs, 1958	F	
Procambarus erythrops Relyea and Sutton, 1975	F	Santa Fe cave crayfish
Procambarus escambiensis Hobbs, 1942	F	
Procambarus evermanni (Faxon, 1890)	F	
Procambarus fallax (Hagen, 1870)	F	
Procambarus fitzpatricki Hobbs, 1972	F	spinytail crayfish
Procambarus franzi Hobbs and Lee, 1976	F	Orange Lake Cave crayfish
Procambarus geminus Hobbs, 1975	F	
Procambarus geodytes Hobbs, 1942	F	
Procambarus gibbus Hobbs, 1969	F	
Procambarus gracilis (Bundy, 1876)	F	prairie crayfish
Procambarus hagenianus (Faxon, 1884)	F	southeastern prairie crayfish
Procambarus hayi (Faxon, 1884)	F	

SCIENTIFIC NAME	OCCURRENCE	COMMON NAME
Procambarus hinei (Ortmann, 1905)	F	
Procambarus hirsutus Hobbs, 1958	F	
Procambarus horsti Hobbs and Means, 1972	F	Big Blue Springs cave crayfish
Procambarus howellae Hobbs, 1952	F	
Procambarus hubbelli (Hobbs, 1940)	F	
Procambarus hybus Hobbs and Walton, 1957	F	
Procambarus incilis Penn, 1962	F	
Procambarus jaculus Hobbs and Walton, 1957	F	javelin crayfish
Procambarus kilbyi (Hobbs, 1940)	F	
Procambarus lagniappe Black, 1968	F	lagniappe crayfish
Procambarus latipleurum Hobbs, 1942	F	
Procambarus lecontei (Hagen, 1870)	F	Mobile crayfish
Procambarus leitheuseri Franz and Hobbs, 1983	F	Coastal Lowland cave crayfish
Procambarus leonensis Hobbs, 1942	F	
Procambarus leptodactylus Hobbs, 1947	F	Pee Dee lotic crayfish
Procambarus lewisi Hobbs and Walton, 1959	F	
Procambarus liberorum Fitzpatrick, 1978	F	Osage burrowing crayfish
Procambarus litosternum Hobbs, 1947	F	
Procambarus lophotus Hobbs and Walton, 1960	F	
Procambarus lucifugus (Hobbs, 1940)	F	Florida cave crayfish
Procambarus lunzi (Hobbs, 1940)	F	
Procambarus lyeli Fitzpatrick and Hobbs, 1971	F	Shutispear crayfish
Procambarus mancus Hobbs and Walton, 1957	F	
Procambarus marthae Hobbs, 1975	F	
Procambarus medialis Hobbs, 1975	F	
Procambarus milleri Hobbs, 1971	F	Miami cave crayfish
Procambarus natchitochae Penn, 1953	F	
Procambarus okaloosae Hobbs, 1942	F	
Procambarus orcinus Hobbs and Means, 1972	F	Woodville Karst cave crayfish
Procambarus ouachitae Penn, 1954	F	
Procambarus paeninsulanus (Faxon, 1914)	F	
Procambarus pallidus (Hobbs, 1940)	F	pallid cave crayfish
Procambarus parasimulans Hobbs and Robinson, 1986	F	
Procambarus pearsei (Creaser, 1934)	F	
Procambarus pecki Hobbs, 1967	F	phantom cave crayfish
Procambarus penni Hobbs, 1951	F	Pearl blackwater crayfish
Procambarus petersi Hobbs, 1981	F	
Procambarus pictus (Hobbs, 1940)	F	spotted royal crayfish
Procambarus planirostris Penn, 1953	F	
Procambarus plumimanus Hobbs and Walton, 1958	F	
Procambarus pogon Fitzpatrick, 1978	F	
Procambarus pubescens (Faxon, 1884)	F	bearded red crayfish
Procambarus pubischelae Hobbs, 1942	F	
Procambarus pycnogonopodus Hobbs, 1942	F	
Procambarus pygmaeus Hobbs, 1942	F	
Procambarus raneyi Hobbs, 1953	F	Christmas tree crayfish
Procambarus rathbunae (Hobbs, 1940)	F	
Procambarus reimeri Hobbs, 1979	F	
Procambarus rogersi (Hobbs, 1938)	F	
Procambarus seminolae Hobbs, 1942	F	

SCIENTIFIC NAME	OCCURRENCE	COMMON NAME
Procambarus shermani Hobbs, 1942	F	
Procambarus simulans (Faxon, 1884)	F	
Procambarus spiculifer (Le Conte, 1856)	F	
Procambarus suttkusi Hobbs, 1953	F	
Procambarus talpoides Hobbs, 1981	F	
Procambarus tenuis Hobbs, 1950	F	
Procambarus texanus Hobbs, 1971	F	
Procambarus troglodytes (Le Conte, 1856)	F	
Procambarus truculentus Hobbs, 1954	F	
Procambarus tulanei Penn, 1953	F	
Procambarus verrucosus Hobbs, 1952	F	
Procambarus versutus (Hagen, 1870)	F	
Procambarus viaeviridis (Faxon, 1914)	F	
Procambarus vioscai Penn, 1946	F	
Procambarus youngi Hobbs, 1942	F	Florida longbeak crayfish
Troglocambarus maclanei Hobbs, 1942	F	spider cave crayfish

Superfamily Enoplometopoidea

Enoplometopidae

Enoplometopus antillensis (Lütken, 1865)	A	flaming reef lobster

INFRAORDER THALASSINIDEA

Superfamily Thalassinoidea

Axiidae — lobster shrimps

Axiopsis hirsutimana (Boesch and Smalley, 1972)	A	
Axiopsis jenneri (Williams, 1974)	A	
Axiopsis oxypleura (Williams, 1974)	A	
Axiopsis spinulicauda (Rathbun, 1902)	P	
Axius agassizi Bouvier, 1905	A	
Axius armatus Smith, 1881	A	
Axius borradailei Bouvier, 1905	A	
Axius communis Bouvier, 1905	A	
Axius rotundifrons Bouvier, 1905	A	
Axius serratus Stimpson, 1852	A	
Calastacus investigatoris Anderson, 1896	P	
Calastacus quinqueseriatus Rathbun, 1902	P	
Calastacus stilirostris Faxon, 1893	P	
Calocaris templemanni Squires, 1965	A	
Coralaxius abelei Kensley and Gore, 1981	A	

Callianassidae—ghost shrimps

Callianassa affinis Holmes, 1900	P	tidepool ghost shrimp
Callianassa biformis Biffar, 1971	A	biform ghost shrimp
Callianassa branneri (Rathbun, 1900)	A	bighand ghost shrimp
Callianassa californiensis Dana, 1854	P	bay ghost shrimp
Callianassa fragilis Biffar, 1970	A–P	fragile ghost shrimp
Callianassa gigas Dana, 1852	P	giant ghost shrimp

SCIENTIFIC NAME	OCCURRENCE	COMMON NAME
Callianassa goniophthalma Rathbun, 1902	P	slope ghost shrimp
Callianassa guassutinga Rodrigues, 1966	A	
Callianassa jamaicensis Schmitt, 1935	A	estuarine ghost shrimp
Callianassa louisianensis (Schmitt, 1935)	A	
Callianassa marginata Rathbun, 1901	A	
Callianassa rathbunae Schmitt, 1935	A	
Callianassa setimanus (De Kay, 1926)	A	
Callianassa trilobata Biffar, 1970	A	
Calliax quadracuta (Biffar, 1970)	A	
Callichirus islagrande (Schmitt, 1935)	A	beach ghost shrimp
Callichirus major (Say, 1818)	A	Carolinian ghost shrimp
Corallianassa longiventris (A. Milne Edwards, 1870)	A	
Ctenocheles leviceps Rabalais, 1979	A	
Glypturus acanthochirus (Stimpson, 1866)	A	
Gourretia latispina (Dawson, 1967)	A	broadspine ghost shrimp

Laomediidae

Naushonia crangonoides Kingsley, 1897	A	

Upogebiidae—mud shrimps

Upogebia affinis (Say, 1818)	A	coastal mud shrimp
Upogebia lepta Williams, 1986	P	
Upogebia macginitieorum Williams, 1986	P	
Upogebia onychion Williams, 1986	P	
Upogebia operculata Schmitt, 1924	A	operculate mud shrimp
Upogebia pugettensis (Dana, 1852)	P	blue mud shrimp

INFRAORDER PALINURA

Superfamily Eryonoidea

Polychelidae

Polycheles crucifer (Willemoes-Suhm, 1873)	A	
Polycheles granulatus Faxon, 1893	A	
Polycheles typhlops Heller, 1862	A	
Polycheles validus (A. Milne Edwards, 1880)	A	
Stereomastis nana (Smith, 1884)	A–P	
Stereomastis sculpta (Smith, 1880)	A–P	
Willemoesia forceps A. Milne Edwards, 1880	A	

Superfamily Palinuroidea

Palinuridae—spiny lobsters

Justitia longimanus (H. Milne Edwards, 1837)	A	
Panulirus argus (Latreille, 1804)	A	Caribbean spiny lobster
Panulirus guttatus (Latreille, 1804)	A	spotted spiny lobster
Panulirus interruptus Randall, 1839	P	California spiny lobster
Panulirus laevicauda (Latreille, 1817)	A	smoothtail spiny lobster

SCIENTIFIC NAME	OCCURRENCE	COMMON NAME

Scyllaridae—slipper lobsters

Parribacus antarcticus (Lund, 1793)	A–P	...... sculptured slipper lobster
Scyllarides aequinoctialis (Lund, 1793)	A–P	.... Spanish slipper lobster
Scyllarides nodifer (Stimpson, 1866)	A	.... ridged slipper lobster
Scyllarus americanus (Smith, 1869)	A	.... American slipper lobster
Scyllarus chacei Holthuis, 1960	A	.... Chace slipper lobster
Scyllarus depressus (Smith, 1881)	A	.... scaled slipper lobster

Synaxiidae—furry lobsters

Palinurellus gundlachi (von Martens, 1878)	A	.... furry lobster

INFRAORDER ANOMURA

Superfamily Paguroidea

Pomatochelidae

Pylocheles scutata Ortmann, 1892	A	....

Coenobitidae—land hermit crabs

Coenobita clypeatus (Herbst, 1791)	A	.... land hermit

Diogenidae—left-handed hermit crabs

Calcinus tibicen (Herbst, 1791)	A	.... orangeclaw hermit
Cancellus ornatus Benedict, 1901	A	....
Cancellus viridis Mayo, 1973	A	....
Clibanarius antillensis Stimpson, 1862	A	....
Clibanarius cubensis (de Saussure, 1858)	A	.... widestripe hermit
Clibanarius tricolor (Gibbes, 1850)	A	.... tricolor hermit
Clibanarius vittatus (Bosc, 1802)	A	.... thinstripe hermit
Dardanus fucosus Biffar and Provenzano, 1972	A	.... bareye hermit
Dardanus insignis (de Saussure, 1858)	A	.... red brocade hermit
Dardanus venosus (H. Milne Edwards, 1848)	A	.... stareye hermit
Isocheles pilosus (Holmes, 1900)	P	.... moon snail hermit
Isocheles wurdemanni Stimpson, 1862	A	.... surf hermit
Paguristes anomalus Bouvier, 1918	A	....
Paguristes bakeri Holmes, 1900	P	.... digger hermit
Paguristes cadenati Forest, 1954	A	.... red reef hermit
Paguristes erythrops Holthuis, 1959	A	....
Paguristes grayi Benedict, 1901	A	....
Paguristes hernancortezi McLaughlin and Provenzano, 1974	A	....
Paguristes hewatti Wass, 1963	A	....
Paguristes hummi Wass, 1955	A	....
Paguristes inconstans McLaughlin and Provenzano, 1974	A	....
Paguristes invisisacculus McLaughlin and Provenzano, 1974	A	....

SCIENTIFIC NAME	OCCURRENCE	COMMON NAME
Paguristes laticlavus McLaughlin and Provenzano, 1974	A	
Paguristes limonensis McLaughlin and Provenzano, 1974	A	
Paguristes lymani A. Milne Edwards and Bouvier, 1893	A	
Paguristes moorei Benedict, 1901	A	
Paguristes oxyophthalmus Holthuis, 1959	A	
Paguristes parvus Holmes, 1900	P	island hermit
Paguristes puncticeps Benedict, 1901	A	
Paguristes sericeus A. Milne Edwards, 1880	A	blue-eye hermit
Paguristes spinipes A. Milne Edwards, 1880	A	
Paguristes starki Provenzano, 1965	A	
Paguristes tortugae Schmitt, 1933	A	bandeye hermit
Paguristes triangulatus A. Milne Edwards and Bouvier, 1893	A	
Paguristes turgidus (Stimpson, 1856)	P	
Paguristes ulreyi Schmitt, 1921	P	furry hermit
Paguristes wassi Provenzano, 1961	A	
Petrochirus diogenes (Linnaeus, 1758)	A	giant hermit

Paguridae—right-handed hermit crabs

Agaricochirus acanthinus McLaughlin, 1982	A	
Agaricochirus alexandri (A. Milne Edwards and Bouvier, 1893)	A	
Agaricochirus boletifer (A. Milne Edwards and Bouvier, 1893)	A	
Agaricochirus gibbosimanus (A. Milne Edwards, 1880)	A	
Anisopagurus bartletti (A. Milne Edwards, 1880)	A	
Anisopagurus pygmaeus (Bouvier, 1918)	A	
Catapaguroides microps A. Milne Edwards and Bouvier, 1893	A	
Catapagurus gracilis (Smith, 1880)	A	
Catapagurus sharreri A. Milne Edwards, 1880	A	
Discorsopagurus schmitti (Stevens, 1925)	P	
Elassochirus cavimanus (Miers, 1879)	P	purple hermit
Elassochirus gilli (Benedict, 1892)	P	Pacific red hermit
Elassochirus tenuimanus (Dana, 1851)	P	wideband hermit
Enallopaguropsis guatemoci (Glassell, 1937)	P	
Haigia diegensis (Scanland and Hopkins, 1969)	P	
Iridopagurus caribbensis (A. Milne Edwards and Bouvier, 1893)	A	
Iridopagurus globulus de Saint Laurent-Dechancé, 1966	A	
Iridopagurus iris (A. Milne Edwards, 1880)	A	
Iridopagurus reticulatus García-Gómez, 1983	A	
Iridopagurus violaceus de Saint Laurent-Dechancé, 1966	A	
Labidochirus splendescens (Owen, 1839)	P	splendid hermit
Manucomplanus corallinus (Benedict, 1892)	A	

SCIENTIFIC NAME	OCCURRENCE	COMMON NAME
Nematopaguroides cf. *fagei*		
Forest and de Saint Laurent, 1967	A	....
Nematopaguroides pusillus		
Forest and de Saint Laurent, 1967	A	....
Ostraconotus spatulipes (A. Milne Edwards, 1880) ...	A	....
Orthopagurus minimus (Holmes, 1900)	P	....
Pagurus acadianus Benedict, 1901	A	....
Pagurus aleuticus (Benedict, 1892)	P	.... Aleutian hermit
Pagurus annulipes (Stimpson, 1860)	A	....
Pagurus arcuatus Squires, 1964	A	....
Pagurus armatus (Dana, 1851)	P	.... armed hermit
Pagurus beringanus (Benedict, 1892)	P	....
Pagurus bouvieri Faxon, 1895	A	....
Pagurus brandti (Benedict, 1892)	P	.... sponge hermit
Pagurus brevidactylus (Stimpson, 1859)	A	....
Pagurus bullisi Wass, 1963	A	....
Pagurus capillatus (Benedict, 1892)	P	....
Pagurus carolinensis McLaughlin, 1975	A	.... wormreef hermit
Pagurus caurinus Hart, 1971	P	.... greenmark hermit
Pagurus confragosus (Benedict, 1892)	P	.... knobbyhand hermit
Pagurus cornutus (Benedict, 1892)	P	.... hornyhand hermit
Pagurus criniticornis (Dana, 1852)	A	....
Pagurus curacaoensis (Benedict, 1892)	A	....
Pagurus dalli (Benedict, 1892)	P	.... whiteknee hermit
Pagurus defensus (Benedict, 1892)	A	....
Pagurus dissimilis		
(A. Milne Edwards and Bouvier, 1893)	A	....
Pagurus granosimanus (Stimpson, 1859)	P	.... grainyhand hermit
Pagurus gymnodactylus Lemaitre, 1982	A	....
Pagurus hemphilli (Benedict, 1892)	P	.... maroon hermit
Pagurus hirsutiusculus (Dana, 1851)	P	.... hairy hermit
Pagurus impressus (Benedict, 1892)	A	.... dimpled hermit
Pagurus kennerlyi (Stimpson, 1864)	P	.... bluespine hermit
Pagurus longicarpus Say, 1817	A	.... longwrist hermit
Pagurus maclaughlinae García-Gómez, 1982	A	....
Pagurus marshi Benedict, 1901	A	....
Pagurus mertensii Brandt, 1851	P	....
Pagurus middendorfii Brandt, 1851	P	....
Pagurus ochotensis Brandt,1851	P	.... Alaskan hermit
Pagurus piercei Wass, 1963	A	.... coralthicket hermit
Pagurus politus (Smith, 1882)	A	....
Pagurus pollicaris Say, 1817	A	.... flatclaw hermit
Pagurus provenzanoi		
Forest and de Saint Laurent, 1967	A	....
Pagurus pubescens Krøyer, 1838	A	....
Pagurus quaylei Hart, 1971	P	....
Pagurus rathbuni (Benedict, 1892)	P	.... longfinger hermit
Pagurus redondoensis Wicksten, 1982	P	.... bandclaw hermit
Pagurus rotundimanus Wass, 1963	A	....
Pagurus samuelis (Stimpson, 1857)	P	.... blueband hermit
Pagurus setosus (Benedict, 1892)	P	.... setose hermit

SCIENTIFIC NAME	OCCURRENCE	COMMON NAME
Pagurus spilocarpus Haig, 1977	P	.... spotwrist hermit
Pagurus stevensae Hart, 1971	P	.... Stevens hermit
Pagurus stimpsoni		
(A. Milne Edwards and Bouvier, 1893)	A	....
Pagurus tanneri (Benedict, 1892)	P	.... longhand hermit
Pagurus townsendi (Benedict, 1892)	P	....
Pagurus trigonocheirus (Stimpson, 1858)	P	.... fuzzy hermit
Pagurus undosus (Benedict, 1892)	P	.... Pribilof hermit
Parapagurodes laurentae		
McLaughlin and Haig, 1973	P	....
Parapagurodes makarovi		
McLaughlin and Haig, 1973	P	....
Phimochirus californiensis (Benedict, 1892)	P	....
Phimochirus holthuisi (Provenzano, 1961)	A	.... red-striped hermit
Phimochirus leurocarpus McLaughlin, 1981	A	....
Phimochirus operculatus (Stimpson, 1859)	A	.... polkadotted hermit
Phimochirus randalli (Provenzano, 1961)	A	....
Pylopaguropsis atlantica Wass, 1963	A	....
Pylopagurus discoidalis (A. Milne Edwards, 1880) ...	A	....
Pylopagurus holmesi Schmitt, 1921	P	....
Rhodochirus rosaceus		
(A. Milne Edwards and Bouvier, 1893)	A	.... stellate hermit
Solenopagurus lineatus (Wass, 1963)	A	....
Tomopaguropsis problematica		
(A. Milne Edwards and Bouvier, 1893)	A	....
Tomopagurus chacei (Wass, 1963)	A	....
Tomopagurus cokeri (Hay, 1917)	A	....
Tomopagurus cubensis (Wass, 1963)	A	....
Tomopagurus rubropunctatus		
A. Milne Edwards and Bouvier, 1893	A	....
Tomopagurus wassi McLaughlin, 1981	A	....

Parapaguridae—deepwater hermit crabs

Parapagurus abyssorum Filhol, 1855	A	....
Parapagurus alaminos Lemaitre, 1986	A	....
Parapagurus arcuatus		
(A. Milne Edwards and Bouvier, 1893)	A	....
Parapagurus bicristatus (A. Milne Edwards, 1880) ...	A	....
Parapagurus haigae de Saint Laurent, 1972	P	....
Parapagurus nudus (A. Milne Edwards, 1891)	A	....
Parapagurus pictus (Smith, 1883)	A	....
Parapagurus pilimanus (A. Milne Edwards, 1880) ...	P	....
Parapagurus pilosimanus Smith, 1879	A–P	....

Lithodidae—stone and king crabs

Acantholithodes hispidus (Stimpson, 1860)	P	....
Cryptolithodes sitchensis Brandt, 1853	P	.... umbrella crab
Cryptolithodes typicus Brandt, 1849	P	.... butterfly crab
Dermaturus mandtii Brandt, 1849	P	....
Glyptolithodes cristatipes (Faxon, 1893)	P	....

SCIENTIFIC NAME	OCCURRENCE	COMMON NAME
Hapalogaster cavicauda Stimpson, 1859	P	.... furry crab
Hapalogaster grebnitzkii Schalfeew, 1892	P	....
Hapalogaster mertensii Brandt, 1849	P	.... hairy crab
Lithodes aequispina Benedict, 1894	P	.... golden king crab
Lithodes couesi Benedict, 1894	P	.... scarlet king crab
Lithodes maja (Linnaeus, 1758)	A	....
Lopholithodes foraminatus (Stimpson, 1859)	P	....
Lopholithodes mandtii Brandt, 1849	P	....
Neolithodes agassizii (Smith, 1882)	A	....
Neolithodes grimaldii (A. Milne Edwards and Bouvier, 1894)	A	....
Oedignathus inermis (Stimpson, 1860)	P	....
Paralithodes californiensis (Benedict, 1894)	P	.... California king crab
Paralithodes camtschaticus (Tilesius, 1815)	P	.... red king crab
Paralithodes platypus Brandt, 1850	P	.... blue king crab
Paralithodes rathbuni (Benedict, 1894)	P	....
Paralomis cubensis Chace, 1939	A	....
Paralomis multispina (Benedict, 1894)	P	....
Paralomis verrilli (Benedict, 1894)	P	....
Phyllolithodes papillosus Brandt, 1849	P	.... flatspine triangle crab
Placetron wosnessenskii Schalfeew, 1892	P	.... scaled crab
Rhinolithodes wosnessenskii Brandt, 1849	P	.... rhinoceros crab

Superfamily Galatheoidea

Chirostylidae

Chirostylus perarmatus Haig, 1968	P	....
Eumunida picta Smith, 1883	A	....
Uroptychus brevis Benedict, 1902	A	....
Uroptychus armatus (A. Milne Edwards, 1880)	A	....
Uroptychus nitidus (A. Milne Edwards, 1880)	A	....

Galatheidae—squat lobsters

Galathea californiensis (Benedict, 1902)	P	....
Galathea rostrata A. Milne Edwards, 1880	A	....
Munida affinis A. Milne Edwards, 1880	A	....
Munida angulata Benedict, 1902	A	....
Munida flinti Benedict, 1902	A	....
Munida forceps A. Milne Edwards, 1880	A	....
Munida hispida Benedict, 1902	P	....
Munida iris A. Milne Edwards, 1880	A	....
Munida irrasa A. Milne Edwards, 1880	A	....
Munida longipes A. Milne Edwards, 1880	A	....
Munida microphthalma A. Milne Edwards, 1880	A	....
Munida miles A. Milne Edwards, 1880	A	....
Munida nuda Benedict, 1902, ...	A	....
Munida pusilla Benedict, 1902	A	....
Munida quadrispina Benedict, 1902	P	....
Munida sanctipauli Henderson, 1885	A	....
Munida simplex Benedict, 1902	A	....

SCIENTIFIC NAME	OCCURRENCE	COMMON NAME
Munida spinifrons Henderson, 1885	A	
Munida stimpsoni A. Milne Edwards, 1880	A	
Munida tenuimana G. O. Sars, 1871	A	
Munida valida Smith, 1883	A	
Munidopsis abbreviata (A. Milne Edwards, 1880)	A	
Munidopsis abdominalis (A. Milne Edwards, 1880)	A	
Munidopsis acuminata Benedict, 1902	A	
Munidopsis alaminos Pequegnat and Pequegnat, 1970	A	
Munidopsis albatrossae Pequegnat and Pequegnat, 1973	P	
Munidopsis aries (A. Milne Edwards, 1880)	A	
Munidopsis armata (A. Milne Edwards, 1880)	A	
Munidopsis aspera (Henderson, 1885)	P	
Munidopsis bairdii (Smith, 1884)	A–P	
Munidopsis barbarae (Boone, 1927)	A	
Munidopsis beringana Benedict, 1902	P	
Munidopsis bermudezi Chace, 1939	A	
Munidopsis cascadia Ambler, 1980	P	
Munidopsis ciliata Wood-Mason, 1891	P	
Munidopsis crassa Smith, 1885	A	
Munidopsis cubensis Chace, 1942	A	
Munidopsis curvirostra Whiteaves, 1874	A	
Munidopsis depressa Faxon, 1893	P	
Munidopsis diomedeae (Faxon, 1893)	P	
Munidopsis erinacea (A. Milne Edwards, 1880)	A	
Munidopsis expansa Benedict, 1902	A	
Munidopsis gilli Benedict, 1902	A	
Munidopsis gulfensis Pequegnat and Pequegnat, 1971	A	
Munidopsis hystrix Faxon, 1893	P	
Munidopsis latifrons (A. Milne Edwards, 1880)	A	
Munidopsis latirostris Faxon, 1893	P	
Munidopsis livida (A. Milne Edwards, 1886)	A	
Munidopsis longimanus (A. Milne Edwards, 1880)	A	
Munidopsis pallida Alcock, 1894	P	
Munidopsis palmata Khodkina, 1975	P	
Munidopsis platirostris (A. Milne Edwards and Bouvier, 1894)	A	
Munidopsis polita (Smith, 1883)	A	
Munidopsis quadrata Faxon, 1893	P	
Munidopsis robusta (A. Milne Edwards, 1880)	A	
Munidopsis rostrata (A. Milne Edwards, 1880)	A	
Munidopsis scabra Faxon, 1893	P	
Munidopsis serratifrons (A. Milne Edwards, 1880)	A	
Munidopsis serricornis (Loven, 1852)	A	
Munidopsis sigsbei (A. Milne Edwards, 1880)	A	
Munidopsis similis Smith, 1885	A	
Munidopsis simplex (A. Milne Edwards, 1880)	A	
Munidopsis spinifera (A. Milne Edwards, 1880)	A	
Munidopsis spinoculata (A. Milne Edwards, 1880)	A	
Munidopsis spinosa (A. Milne Edwards, 1880)	A	

SCIENTIFIC NAME	OCCURRENCE	COMMON NAME
Munidopsis transtridens Pequegnat and Pequegnat, 1971	A	
Munidopsis tuftsi Ambler, 1980	P	
Munidopsis verrilli Benedict, 1980	P	
Munidopsis verrucosa Khodkina, 1975	P	
Munidopsis yaquinensis Ambler, 1980	P	
Pleuroncodes planipes Stimpson, 1860	P	pelagic red crab

Porcellanidae—porcelain crabs

SCIENTIFIC NAME	OCCURRENCE	COMMON NAME
Euceramus praelongus Stimpson, 1860	A	olivepit porcelain crab
Megalobrachium poeyi (Guérin-Méneville, 1855)	A	hairyclaw porcelain crab
Megalobrachium soriatum (Say, 1818)	A	pentagonal porcelain crab
Neopisosoma angustifrons (Benedict, 1901)	A	
Pachycheles ackleianus A. Milne Edwards, 1880	A	red-reef porcelain crab
Pachycheles holosericus Schmitt, 1921	P	sponge porcelain crab
Pachycheles monilifer (Dana, 1852)	A	wormreef porcelain crab
Pachycheles pilosus (H. Milne Edwards, 1837)	A	pilose porcelain crab
Pachycheles pubescens Holmes, 1900	P	pubescent porcelain crab
Pachycheles riisei (Stimpson, 1858)	A	Riise porcelain crab
Pachycheles rudis Stimpson, 1859	P	thickclaw porcelain crab
Pachycheles rugimanus A. Milne Edwards, 1880	A	sculptured porcelain crab
Parapetrolisthes tortugensis (Glassell, 1945)	A	spiny porcelain crab
Petrolisthes armatus (Gibbes, 1850)	A	green porcelain crab
Petrolisthes cabrilloi Glassell, 1945	P	Cabrillo porcelain crab
Petrolisthes cinctipes (Randall, 1839)	P	flat porcelain crab
Petrolisthes eriomerus Stimpson, 1871	P	flattop crab
Petrolisthes galathinus (Bosc, 1802)	A	banded porcelain crab
Petrolisthes jugosus Streets, 1872	A	red-white porcelain crab
Petrolisthes manimaculis Glassell, 1945	P	chocolate porcelain crab
Petrolisthes politus (Gray, 1831)	A	redback porcelain crab
Petrolisthes rathbunae Schmitt, 1916	P	Rathbun porcelain crab
Polyonyx gibbesi Haig, 1956	A	eastern tube crab
Polyonyx quadriungulatus Glassell, 1935	P	western tube crab
Porcellana sayana (Leach, 1820)	A	spotted porcelain crab
Porcellana sigsbeiana A. Milne Edwards, 1880	A	striped porcelain crab
Porcellana stimpsoni A. Milne Edwards, 1880	A	Stimpson porcelain crab

Superfamily Hippoidea

Albuneidae—mole crabs

SCIENTIFIC NAME	OCCURRENCE	COMMON NAME
Albunea gibbesii Stimpson, 1859	A	surf mole crab
Albunea paretii Guérin-Méneville, 1853	A	beach mole crab
Blepharipoda occidentalis Randall, 1839	P	spiny mole crab
Lepidopa benedicti Schmitt, 1935	A	
Lepidopa californica Efford, 1971	P	California mole crab
Lepidopa websteri Benedict, 1903	A	
Lophomastix diomedeae Benedict, 1904	P	
Zygopa michaelis Holthuis, 1960	A	blind mole crab

SCIENTIFIC NAME	OCCURRENCE	COMMON NAME
Hippidae—sand crabs		
Emerita analoga (Stimpson, 1857)	P	Pacific sand crab
Emerita benedicti Schmitt, 1935	A	Benedict sand crab
Emerita portoricensis Schmitt, 1935	A	Puerto Rican sand crab
Emerita talpoida (Say, 1857)	A	Atlantic sand crab
Hippa cubensis (Saussure, 1857)	A	

INFRAORDER BRACHYURA

SECTION DROMIACEA

Superfamily Dromioidea

Dromiidae—sponge crabs		
Dromia erythropus (George Edwards, 1771)	A	redeye sponge crab
Dromidia antillensis Stimpson, 1858	A	hairy sponge crab
Dromidia larraburei Rathbun, 1910	P	
Hypoconcha arcuata Stimpson, 1858	A	granulate shellback crab
Hypoconcha californiensis Bouvier, 1898	P	California shellback crab
Hypoconcha parasitica (Linnaeus, 1763)	A	rough shellback crab
Hypoconcha spinosissima Rathbun, 1933	A	spiny shellback crab

Homolodromiidae		
Dicranodromia ovata A. Milne Edwards, 1880	A	

SECTION ARCHAEOBRACHYURIA

Superfamily Cyclodorippoidea

Cyclodorippidae		
Clythrocerus decorus Rathbun, 1933	P	
Clythrocerus granulatus Rathbun, 1898	A	
Clythrocerus nitidus (A. Milne Edwards, 1880)	A	
Clythrocerus perpusillus Rathbun, 1901	A	
Clythrocerus planus (Rathbun, 1900)	P	chip crab
Clythrocerus stimpsoni Rathbun, 1937	A	
Cyclodorippe antennaria A. Milne Edwards, 1880	A	

Cymonomidae		
Cymonomus quadratus A. Milne Edwards, 1880	A	
Cymopolus agassizi A. Milne Edwards and Bouvier, 1899	A	

Superfamily Homoloidea

Homolidae—carrier crabs		
Homola barbata (Fabricius, 1793)	A	
Homola vigil A. Milne Edwards, 1880	A	
Hypsophrys noar Williams, 1974	A	

SCIENTIFIC NAME	OCCURRENCE	COMMON NAME
Paromola faxoni (Schmitt, 1921)	P	....

Latreilliidae

Latreillia manningi Williams, 1982	A	.... daddy-longlegs crab

Superfamily Raninoidea

Raninidae—frog crabs

Lyreidus nitidus (A. Milne Edwards, 1880)	A	.... longleg frog crab
Ranilia constricta (A. Milne Edwards, 1880)	A	.... sicklefoot frog crab
Ranilia muricata H. Milne Edwards, 1837	A	.... muricate frog crab
Raninoides loevis (Latreille, 1825)	A	.... furrowed frog crab
Raninoides louisianensis Rathbun, 1933	A	.... gulf frog crab
Symethis variolosa (Fabricius, 1793)	A	.... eroded frog crab

Section Oxystomata

Superfamily Dorippoidea

Dorippidae—sumo crabs

Ethusa mascarone americana A. Milne Edwards, 1880	A	.... stalkeye sumo crab
Ethusa microphthalma Smith, 1881	A	.... broadback sumo crab
Ethusa tenuipes Rathbun, 1897	A	.... spiketoe sumo crab
Ethusa truncata A. Milne Edwards and Bouvier, 1899 ...	A	.... truncate sumo crab
Ethusina abyssicola Smith, 1884	A	.... abyssal sumo crab

Superfamily Leucosioidea

Calappidae—box crabs

Acanthocarpus alexandri Stimpson, 1871	A	.... gladiator box crab
Acanthocarpus bispinosus A. Milne Edwards, 1880 ...	A	.... twospine box crab
Calappa angusta A. Milne Edwards, 1880	A	.... nodose box crab
Calappa flammea (Herbst, 1794)	A	.... flame box crab
Calappa gallus (Herbst, 1803)	A	.... rough box crab
Calappa ocellata Holthuis, 1958	A	.... ocellate box crab
Calappa sulcata Rathbun, 1898	A	.... yellow box crab
Cycloes bairdii Stimpson, 1860	A	.... shameface heart crab
Hepatus epheliticus (Linnaeus, 1763)	A	.... calico box crab
Hepatus pudibundus (Herbst, 1785)	A	.... flecked box crab
Mursia gaudichaudii (H. Milne Edwards, 1837)	P	.... armed box crab
Osachila antillensis Rathbun, 1893	A	....
Osachila semilevis Rathbun, 1916	A	.... thinlip jewelbox crab
Osachila tuberosa Stimpson, 1871	A	.... thicklip jewelbox crab

Leucosiidae—purse crabs

Callidactylus asper Stimpson, 1871	A	.... spurfinger purse crab
Ebalia cariosa (Stimpson, 1860)	A	.... sculptured clutch crab

SCIENTIFIC NAME	OCCURRENCE	COMMON NAME
Ebalia stimpsonii A. Milne Edwards, 1880	A	.... thinarm clutch crab
Iliacantha intermedia Miers, 1886	A	.... granulose purse crab
Iliacantha liodactylus Rathbun, 1898	A	....
Iliacantha sparsa Stimpson, 1871	A	.... shouldered purse crab
Iliacantha subglobosa Stimpson, 1871	A	.... longfinger purse crab
Lithadia cadaverosa Stimpson, 1871	A	.... carinate clutch crab
Lithadia granulosa A. Milne Edwards, 1880	A	....
Myropsis quinquespinosa Stimpson, 1871	A	.... fivespine purse crab
Persephona crinita Rathbun, 1931	A	.... pink purse crab
Persephona mediterranea (Herbst, 1794)	A	.... mottled purse crab
Randallia bulligera Rathbun, 1898	P	....
Randallia ornata (Randall, 1839)	P	.... globose sand crab
Speloeophorus elevatus Rathbun, 1898	A	.... bighole clutch crab
Speloeophorus nodosus (Bell, 1855)	A	.... twohole clutch crab
Speloeophorus pontifer (Stimpson, 1871)	A	.... smallhole clutch crab
Uhlias limbatus Stimpson, 1871	A	.... eroded clutch crab

SECTION OXYRHYNCA

Superfamily Majoidea

Majidae—spider crabs

Acanthonyx petiverii H. Milne Edwards, 1834	A	.... jacknife spider crab
Aepinus septemspinosus (A. Milne Edwards, 1879) ...	A	....
Anasimus latus Rathbun, 1894	A	.... stilt spider crab
Anomalothir furcillatus (Stimpson, 1871)	A	....
Arachnopsis filipes Stimpson, 1871	A	....
Batrachonotus fragosus Stimpson, 1871	A	....
Chionoecetes angulatus Rathbun, 1924	P	.... triangle Tanner crab
Chionoecetes bairdi Rathbun, 1924	P	.... Tanner crab
Chionoecetes opilio (Fabricius, 1788)	A–P	.... snow crab
Chionoecetes tanneri Rathbun, 1893	P	.... grooved Tanner crab
Chlorilia longipes Dana, 1851	P	.... longhorn decorator crab
Chorinus heros (Herbst, 1790)	A	.... shorthorn decorator crab
Coelocerus spinosus A. Milne Edwards, 1875	A	.... channelnose spider crab
Collodes leptocheles Rathbun, 1894	A	....
Collodes nudus Stimpson, 1871	A	....
Collodes obesus A. Milne Edwards, 1878	A	....
Collodes robustus Smith, 1881	A	....
Collodes trispinosus Stimpson, 1871	A	....
Dorhynchus thomsoni Thomson, 1873	A	....
Epialtoides hiltoni (Rathbun, 1923)	P	.... winged kelp crab
Epialtus bituberculatus H. Milne Edwards, 1834	A	.... variegate spider crab
Epialtus dilatatus A. Milne Edwards, 1878	A	.... winged mime crab
Epialtus kingsleyi Rathbun, 1923	A	....
Epialtus longirostris Stimpson, 1860	A	....
Erileptus spinosus Rathbun, 1893	P	....
Euprognatha gracilipes A. Milne Edwards, 1878	A	....
Euprognatha rastellifera Stimpson, 1871	A	....
Hemus cristulipes A. Milne Edwards, 1875	A	....
Herbstia parvifrons Randall, 1839	P	.... crevice spider crab

SCIENTIFIC NAME	OCCURRENCE	COMMON NAME
Hyas araneus (Linnaeus, 1758)	A	Atlantic lyre crab
Hyas coarctatus Leach, 1815	A–P	Arctic lyre crab
Hyas lyratus Dana, 1815	P	Pacific lyre crab
Inachoides forceps A. Milne Edwards, 1879	A	
Leptopisa setirostris (Stimpson, 1871)	A	
Libinia dubia H. Milne Edwards, 1834	A	longnose spider crab
Libinia emarginata Leach, 1815	A	portly spider crab
Libinia erinacea (A. Milne Edwards, 1879)	A	seagrass spider crab
Loxorhynchus crispatus Stimpson, 1857	P	moss crab
Loxorhynchus grandis Stimpson, 1857	P	sheep crab
Macrocoeloma camptocerum (Stimpson, 1871)	A	Florida decorator crab
Macrocoeloma diplacanthum (Stimpson, 1860)	A	
Macrocoeloma eutheca (Stimpson, 1871)	A	
Macrocoeloma laevigatum (Stimpson, 1860)	A	
Macrocoeloma septemspinosum (Stimpson, 1871)	A	thorny decorator crab
Macrocoeloma subparallelum (Stimpson, 1860)	A	
Macrocoeloma trispinosum (Latreille, 1825)	A	spongy decorator crab
Metoporhaphis calcarata (Say, 1818)	A	false arrow crab
Microphrys antillensis Rathbun, 1920	A	lobed decorator crab
Microphrys bicornuta (Latreille, 1825)	A	speck-claw decorator crab
Mimulus foliatus Stimpson, 1860	P	foliate kelp crab
Mithrax acuticornis Stimpson, 1870	A	sharphorn clinging crab
Mithrax cinctimanus (Stimpson, 1860)	A	banded clinging crab
Mithrax cornutus de Saussure, 1857	A	shorthorn clinging crab
Mithrax coryphe (Herbst, 1801)	A	nodose clinging crab
Mithrax forceps (A. Milne Edwards, 1875)	A	red-ridged clinging crab
Mithrax hemphilli Rathbun, 1892	A	
Mithrax hispidus (Herbst, 1790)	A	coral clinging crab
Mithrax holderi Stimpson, 1871	A	
Mithrax pilosus Rathbun, 1892	A	
Mithrax pleuracanthus Stimpson, 1871	A	shaggy clinging crab
Mithrax ruber (Stimpson, 1871)	A	
Mithrax sculptus (Lamarck, 1818)	A	green clinging crab
Mithrax spinosissimus (Lamarck, 1818)	A	channel clinging crab
Mithrax tortugae Rathbun, 1920	A	
Mithrax verrucosus H. Milne Edwards, 1832	A	paved clinging crab
Mocosoa crebripunctata Stimpson, 1871	A	
Nibilia antilocapra (Stimpson, 1871)	A	shorthorn spiny crab
Oplopisa spinipes A. Milne Edwards, 1879	A	
Oregonia bifurca Rathbun, 1902	P	
Oregonia gracilis Dana, 1851	P	graceful decorator crab
Pelia mutica (Gibbes, 1850)	A	cryptic teardrop crab
Pelia tumida (Lockington, 1877)	P	dwarf teardrop crab
Picroceroides tubularis Miers, 1886	A	
Pitho aculeata (Gibbes, 1850)	A	massive urn crab
Pitho anisodon (von Martens, 1872)	A	oval urn crab
Pitho laevigata (A. Milne Edwards, 1875)	A	eggshell urn crab
Pitho lherminieri (Schramm, 1867)	A	broadback urn crab
Pitho mirabilis (Herbst, 1794)	A	
Pitho quadridentata (Miers, 1879)	A	
Podochela curvirostris (A. Milne Edwards, 1879)	A	

SCIENTIFIC NAME	OCCURRENCE	COMMON NAME
Podochela gracilipes Stimpson, 1871	A	unicorn neck crab
Podochela hemphillii (Lockington, 1877)	P	Hemphill kelp crab
Podochela lamelligera (Stimpson, 1871)	A	
Podochela lobifrons Rathbun, 1925	P	
Podochela macrodera Stimpson, 1860	A	
Podochela riisei Stimpson, 1860	A	longfinger neck crab
Podochela sidneyi Rathbun, 1924	A	shortfinger neck crab
Pugettia dalli Rathbun, 1893	P	spined kelp crab
Pugettia gracilis Dana, 1851	P	graceful kelp crab
Pugettia producta (Randall, 1839)	P	northern kelp crab
Pugettia richii Dana, 1851	P	cryptic kelp crab
Pugettia venetiae Rathbun, 1924	P	Venice kelp crab
Pyromaia arachna Rathbun, 1924	A	needlenose pear crab
Pyromaia cuspidata Stimpson, 1871	A	dartnose pear crab
Pyromaia tuberculata (Lockington, 1876	P	tuberculate pear crab
Rochinia crassa (A. Milne Edwards, 1879)	A	inflated spiny crab
Rochinia hystrix (Stimpson, 1871)	A	quillback spiny crab
Rochinia tanneri (Smith, 1883)	A	thorned spiny crab
Rochinia umbonata (Stimpson, 1871)	A	knobbed spiny crab
Scyra acutifrons Dana, 1851	P	sharpnose crab
Sphenocarcinus corrosus A. Milne Edwards, 1875	A	eroded vase crab
Stenocionops furcatus (Olivier, 1791)	A	furcate spider crab
Stenocionops spinimanus (Rathbun, 1892)	A	prickly spider crab
Stenocionops spinosissimus (Saussure, 1857)	A	tenspine spider crab
Stenorhynchus seticornis (Herbst, 1788)	A	yellowline arrow crab
Stilbomastax margaritifera (Monod, 1939)	A	
Taliepus nuttallii (Randall, 1839)	P	globose kelp crab
Thoe puella Stimpson, 1860	A	scarlet mime crab
Tyche emarginata White, 1847	A	fourhorn crab

Superfamily Parthenopoidea

Parthenopidae—elbow crabs

Cryptopodia concava Stimpson, 1871	A	
Heterocrypta granulata (Gibbes, 1850)	A	smooth elbow crab
Heterocrypta occidentalis (Dana, 1854)	P	sandflat elbow crab
Leiolambrus nitidus Rathbun, 1901	A	white elbow crab
Mesorhoea sexspinosa Stimpson, 1871	A	sixspine elbow crab
Parthenope agona (Stimpson, 1871)	A	yellow elbow crab
Parthenope fraterculus (Stimpson, 1871)	A	rough elbow crab
Parthenope granulata (Kingsley, 1879)	A	bladetooth elbow crab
Parthenope pourtalesii (Stimpson, 1871)	A	spinous elbow crab
Parthenope serrata (H. Milne Edwards, 1834)	A	sawtooth elbow crab
Solenolambrus decemspinosus Rathbun, 1894	A	
Solenolambrus tenellus Stimpson, 1871	A	
Solenolambrus typicus Stimpson, 1871	A	
Tutankhamen cristatipes (A. Milne Edwards, 1880)	A	

SCIENTIFIC NAME	OCCURRENCE	COMMON NAME

SECTION CANCRIDEA

Superfamily Cancroidea

Atelecyclidae—horse crabs

Erimacrus isenbeckii (Brandt, 1848)	P	.... hair crab
Telmessus cheiragonus (Tilesius, 1815)	P	.... helmet crab
Trichopeltarion nobile A. Milne Edwards, 1880	A	.... velvet horse crab

Cancridae—rock crabs

Cancer amphioetus Rathbun, 1898	P	.... bigtooth rock crab
Cancer antennarius Stimpson, 1856	P	.... Pacific rock crab
Cancer anthonyi Rathbun, 1879	P	.... yellow rock crab
Cancer borealis Stimpson, 1859	A	.... Jonah crab
Cancer branneri Rathbun, 1926	P	.... furrowed rock crab
Cancer gracilis Dana, 1852	P	.... graceful rock crab
Cancer irroratus Say, 1817	A	.... Atlantic rock crab
Cancer jordani Rathbun, 1900	P	.... hairy rock crab
Cancer magister Dana, 1852	P	.... Dungeness crab
Cancer oregonensis (Dana, 1852)	P	.... pygmy rock crab
Cancer productus Randall, 1839	P	.... red rock crab

SECTION BRACHYRHYNCHA

Superfamily Portunoidea

Geryonidae—deepsea crabs

Geryon fenneri Manning and Holthuis, 1984	A	.... golden deepsea crab
Geryon quinquedens Smith, 1879	A	.... red deepsea crab

Portunidae—swimming crabs

Arenaeus cribrarius (Lamarck, 1818)	A	.... speckled swimming crab
Bathynectes longispina Stimpson, 1871	A	.... bathyal swimming crab
Benthochascon schmitti Rathbun, 1931	A	...... sharp-oar swimming crab
Callinectes arcuatus Ordway, 1863	P	.... arched swimming crab
Callinectes bellicosus (Stimpson, 1859)	P	.... Cortez swimming crab
Callinectes bocourti A. Milne Edwards, 1897	A	.... Bocourt swimming crab
Callinectes danae Smith, 1869	A	.... Dana swimming crab
Callinectes exasperatus (Gerstaecker, 1856)	A	.... rugose swimming crab
Callinectes larvatus Ordway, 1863	A	.... masked swimming crab
Callinectes ornatus Ordway, 1863	A	.... shelligs
Callinectes rathbunae Contreras, 1930	A	.... sharptoothed swimming crab
Callinectes sapidus Rathbun, 1896	A	.... blue crab
Callinectes similis Williams, 1966	A	.... lesser blue crab
Carcinus maenas (Linnaeus, 1758)	A	.... green crab
Cronius ruber (Lamarck, 1818)	A–P	.... blackpoint sculling crab
Cronius tumidulus (Stimpson, 1871)	A	.... crevice sculling crab
Ovalipes floridanus Hay and Shore, 1918	A	.... Florida lady crab
Ovalipes ocellatus (Herbst, 1799)	A	.... lady crab
Ovalipes stephensoni Williams, 1976	A	.... coarsehand lady crab

SCIENTIFIC NAME	OCCURRENCE	COMMON NAME
Portunus anceps (Saussure, 1858)	A	delicate swimming crab
Portunus binoculus Holthuis, 1969	A	redspotted swimming crab
Portunus depressifrons (Stimpson, 1859)	A	flatface swimming crab
Portunus floridanus Rathbun, 1930	A	
Portunus gibbesii (Stimpson, 1859)	A	iridescent swimming crab
Portunus ordwayi (Stimpson, 1860)	A	redhair swimming crab
Portunus sayi (Gibbes, 1850)	A	sargassum swimming crab
Portunus sebae (H. Milne Edwards, 1834)	A	ocellate swimming crab
Portunus spinicarpus (Stimpson, 1871)	A	longspine swimming crab
Portunus spinimanus Latreille, 1819	A	blotched swimming crab
Portunus ventralis (A. Milne Edwards, 1879)	A	
Portunus vocans (A. Milne Edwards, 1878)	A	
Portunus xantusii (Stimpson, 1860)	P	Xantus swimming crab

Superfamily Xanthoidea

Xanthidae—mud crabs

Actaea acantha (H. Milne Edwards, 1834)	A	spinose rubble crab
Actaea bifrons Rathbun, 1898	A	areolate rubble crab
Allactaea lithostrota Williams, 1974	A	
Banareia palmeri (Rathbun, 1894)	A	hoary rubble crab
Carpilius corallinus (Herbst, 1783)	A	batwing coral crab
Carpoporus papulosus Stimpson, 1871	A	narrowfront rubble crab
Cataleptodius floridanus (Gibbes, 1850)	A	spoonfinger rubble crab
Cataleptodius parvulus (Fabricius, 1793)	A	
Chlorodiella longimana (H. Milne Edwards, 1834)	A	longhand rubble crab
Cycloxanthops novemdentatus (Lockington, 1876)	P	ninetooth pebble crab
Domecia acanthophora (Desbonne and Schramm, 1867)	A	elkhorn coral crab
Dyspanopeus sayi (Smith, 1869)	A	Say mud crab
Dyspanopeus texana (Stimpson, 1859)	A	gulf grassflat crab
Eriphia gonagra (Fabricius, 1781)	A	redfinger rubble crab
Etisus maculatus (Stimpson, 1860)	A	blackfinger rubble crab
Eucratodes agassizii A. Milne Edwards, 1880	A	dirtytooth mud crab
Eurypanopeus abbreviatus (Stimpson, 1860)	A	lobate mud crab
Eurypanopeus depressus (Smith, 1869)	A	flatback mud crab
Eurypanopeus dissimilis (Benedict and Rathbun, 1891)	A	asymmetric mud crab
Eurypanopeus hyperconvexus Garth, 1986	P	
Eurytium limosum (Say, 1818)	A	broadback mud crab
Glyptoxanthus erosus (Stimpson, 1859)	A	eroded mud crab
Gonopanope areolata (Rathbun, 1898)	P	smooth-hand mud crab
Heteractaea ceratopus (Stimpson, 1860)	A	horned mud crab
Heteractaea lunata (H. Milne Edwards and Lucas, 1843)	P	fuzzy mud crab
Hexapanopeus angustifrons (Benedict and Rathbun, 1891)	A	smooth mud crab
Hexapanopeus caribbaeus (Stimpson, 1871)	A	
Hexapanopeus hemphillii (Benedict and Rathbun, 1891)	A	
Hexapanopeus lobipes (A. Milne Edwards, 1880)	A	

SCIENTIFIC NAME	OCCURRENCE	COMMON NAME
Hexapanopeus paulensis Rathbun, 1930	A	.... knobbed mud crab
Hexapanopeus quinquedentatus Rathbun, 1901	A	....
Lobopilumnus agassizii (Stimpson, 1871)	A	.... areolated hairy crab
Lophopanopeus bellus (Stimpson, 1860)	P	.... blackclaw crestleg crab
Lophopanopeus frontalis (Rathbun, 1893)	P	.... molarless crestleg crab
Lophopanopeus leucomanus (Lockington, 1876)	P	.... knobknee crestleg crab
Melybia thalamita Stimpson, 1871	A	.... delicate coral crab
Menippe adina Williams and Felder, 1986	A	.... gulf stone crab
Menippe mercenaria (Say, 1818)	A	.... Florida stone crab
Menippe nodifrons Stimpson, 1859	A	.... Cuban stone crab
Micropanope areolata Rathbun, 1898	P	....
Micropanope barbadensis (Rathbun, 1921)	A	.... prickly mud crab
Micropanope latimanus Stimpson, 1898	P	....
Micropanope lobifrons A. Milne Edwards, 1880	A	.... lobefront mud crab
Micropanope nuttingi (Rathbun, 1898)	A	.... beaded mud crab
Micropanope pusilla A. Milne Edwards, 1880	A	.... puffy mud crab
Micropanope sculptipes Stimpson, 1871	A	.... sculptured mud crab
Micropanope spinipes A. Milne Edwards, 1880	A	.... spiny mud crab
Micropanope urinator (A. Milne Edwards, 1881)	A	.... thorny mud crab
Neopanope packardii (Kingsley, 1871)	A	...... Florida grassflat crab
Panopeus americanus de Saussure, 1857	A	.... narrowback mud crab
Panopeus bermudensis Benedict and Rathbun, 1891 ..	A	.... strongtooth mud crab
Panopeus harttii Smith, 1869	A	.... areolate mud crab
Panopeus herbstii H. Milne Edwards, 1834	A	.... Atlantic mud crab
Panopeus lacustris Desbonne, 1867	A	.... knotfinger mud crab
Panopeus obesus Smith, 1869	A	.... saltmarsh mud crab
Panopeus occidentalis Saussure, 1857	A	.... furrowed mud crab
Panopeus rugosus A. Milne Edwards, 1880	A	.... granulose mud crab
Panopeus simpsoni Rathbun, 1930	A	.... oystershell mud crab
Panopeus turgidus Rathbun, 1930	A	.... ridgeback mud crab
Paractaea rufopunctata nodosa (Stimpson, 1860)	A	.... nodose rubble crab
Paraliomera dispar (Stimpson, 1871)	A	.... black coral crab
Paraliomera longimana (A. Milne Edwards, 1865) ...	A	.... longarm coral crab
Paraxanthias taylori (Stimpson, 1861)	P	.... lumpy rubble crab
Pilumnoides nudifrons (Stimpson, 1871)	A	.... nakedface hairy crab
Pilumnus caribaeus Desbonne and Schramm, 1867 ...	A	.... coarsespined hairy crab
Pilumnus dasypodus Kingsley, 1879	A	.... shortspined hairy crab
Pilumnus floridanus Stimpson, 1871	A	.... plumed hairy crab
Pilumnus gemmatus Stimpson, 1860	A	.... tuberculate hairy crab
Pilumnus holosericus Rathbun, 1898	A	.... roseate hairy crab
Pilumnus lacteus Stimpson, 1871	A	.... velvet hairy crab
Pilumnus longleyi Rathbun, 1930	A	.... studded hairy crab
Pilumnus marshi Rathbun, 1901	A	.... quadrate hairy crab
Pilumnus nudimanus Rathbun, 1901	A	....
Pilumnus pannosus Rathbun, 1896	A	.... beaded hairy crab
Pilumnus sayi Rathbun, 1897	A	.... spineback hairy crab
Pilumnus spinohirsutus (Lockington, 1876)	P	.... retiring hairy crab
Pilumnus spinosissimus Rathbun, 1898	A	.... longspined hairy crab
Platyactaea setigera (H. Milne Edwards, 1834)	A	.... bristled rubble crab
Platypodiella spectabilis (Herbst, 1794)	A	.... gaudy clown crab
Pseudomedaeus agassizii (A. Milne Edwards, 1880) ..	A	.... rough rubble crab

SCIENTIFIC NAME	OCCURRENCE	COMMON NAME
Pseudomedaeus distinctus (Rathbun, 1898)	A	.... armed rubble crab
Rhithropanopeus harrisii (Gould, 1841)	A–P[I]	.... Harris mud crab
Tetraxanthus bidentatus (A. Milne Edwards, 1880) ...	A	.... cornered mud crab
Tetraxanthus rathbunae Chace, 1939	A	.... inflated mud crab
Xanthodius denticulatus (White, 1847)	A	.... denticulate rubble crab

Goneplacidae

Bathyplax typhla A. Milne Edwards, 1880	A	.... rigid bathyal crab
Chacellus filiformis Guinot, 1969	A	.... browed bathyal crab
Chasmocarcinus chacei Felder and Rabalais, 1986 ...	A	....
Chasmocarcinus cylindricus Rathbun, 1901	A	.... smoothwrist soft crab
Chasmocarcinus mississippiensis Rathbun, 1931	A	.... roughwrist soft crab
Eucratopsis crassimanus (Dana, 1852)	A	.... heavyhand rubble crab
Euphrosynoplax clausa Guinot, 1969	A	.... craggy bathyal crab
Euryplax nitida Stimpson, 1859	A	.... glabrous broadface crab
Frevillea barbata A. Milne Edwards, 1880	A	....
Frevillea hirsuta (Borradaile, 1916)	A	.... tufted broadface crab
Glyptoplax smithii A. Milne Edwards, 1880	A	.... truncate rubble crab
Goneplax sigsbei (A. Milne Edwards, 1880)	A	.... dentate broadface crab
Malacoplax californiensis (Lockington, 1877)	P	..... California burrowing crab
Nanoplax xanthiformis (A. Milne Edwards, 1880)	A	.... rough squareback crab
Neopilumnoplax americana (Rathbun, 1898)	A	...... bimarginate bathyal crab
Panoplax depressa Stimpson, 1871	A	.... depressed rubble crab
Pilumnoplax elata (A. Milne Edwards, 1880)	A	....
Prionoplax atlantica Kendall, 1891	A	....
Pseudorhombila quadridentata (Latreille, 1828)	A	.... flecked squareback crab
Robertsella mystica Guinot, 1969	A	.... spiked bathyal crab
Sotoplax robertsi Guinot, 1985	A	....
Speocarcinus carolinensis Stimpson, 1859	A	..... Carolinian squareback crab
Speocarcinus lobatus Guinot, 1969	A	.... gulf squareback crab
Speocarcinus monotuberculatus Felder and Rabalais, 1986	A	....
Thalassoplax angusta Guinot, 1969	A	.... narrow bathyal crab
Trapezioplax tridentata (A. Milne Edwards, 1880) ...	A	.... spined broadface crab

Superfamily Grapsoidea

Gecarcinidae—land crabs

Cardisoma guanhumi Latreille, 1825	A	.... blue land crab
Gecarcinus lateralis (Freminville, 1835)	A	.... blackback land crab
Gecarcinus ruricola (Linnaeus, 1758)	A	.... purple land crab

Grapsidae—shore, marsh, and talon crabs

Aratus pisonii (H. Milne Edwards, 1837)	A	.... mangrove tree crab
Cyclograpsus integer H. Milne Edwards, 1837	A	.... globose shore crab
Eriocheir sinensis H. Milne Edwards, 1853	A[I]	.... Chinese mitten crab
Euchirograpsus americanus A. Milne Edwards, 1880 ..	A	.... American talon crab
Euchirograpsus antillensis Türkay, 1975	A	.... Antillean talon crab
Geograpsus lividus (H. Milne Edwards, 1837)	A	.... variegate shore crab

SCIENTIFIC NAME	OCCURRENCE	COMMON NAME
Goniopsis cruentata (Latreille, 1802)	A	mangrove root crab
Grapsodius eximius Holmes, 1900	P	
Grapsus grapsus (Linnaeus, 1758)	A	Sally Lightfoot crab
Hemigrapsus nudus (Dana, 1851)	P	purple shore crab
Hemigrapsus oregonensis (Dana, 1851)	P	yellow shore crab
Pachygrapsus crassipes Randall, 1839	P	striped shore crab
Pachygrapsus gracilis (de Saussure, 1858)	A	dark shore crab
Pachygrapsus marinus (Rathbun, 1914)	P	drifter crab
Pachygrapsus transversus (Gibbes, 1850)	A	mottled shore crab
Percnon gibbesi (H. Milne Edwards, 1853)	A	nimble spray crab
Plagusia depressa (Fabricius, 1775)	A	tidal spray crab
Planes cyaneus Dana, 1852	P	flotsam crab
Planes minutus (Linnaeus, 1758)	A	gulfweed crab
Platychirograpsus spectabilis de Man, 1896	A	saber crab
Sesarma benedicti Rathbun, 1897	A	fatfinger marsh crab
Sesarma cinereum (Bosc, 1802)	A	squareback marsh crab
Sesarma curacaoense de Man, 1892	A	mangrove marsh crab
Sesarma miersii Rathbun, 1897	A	
Sesarma reticulatum (Say, 1817)	A	heavy marsh crab
Sesarma ricordi H. Milne Edwards, 1853	A	humic marsh crab

Superfamily Pinnotheroidea

Pinnotheridae—pea crabs

SCIENTIFIC NAME	OCCURRENCE	COMMON NAME
Dissodactylus alcocki Rathbun, 1918	A	
Dissodactylus borradailei Rathbun, 1918	A	
Dissodactylus crinitichelis Moreira, 1901	A	seabiscuit pea crab
Dissodactylus mellitae (Rathbun, 1900)	A	sand-dollar pea crab
Dissodactylus primitivus Bouvier, 1917	A	
Dissodactylus rugatus Bouvier, 1917	A	
Dissodactylus stebbingi Rathbun, 1918	A	
Fabia byssomiae (Say, 1818)	A	
Fabia canfieldi Rathbun, 1918	P	
Fabia concharum (Rathbun, 1893)	P	smooth mussel crab
Fabia subquadrata Dana, 1851	P	grooved mussel crab
Fabia tellinae Cobb, 1973	A	
Opisthopus transversus Rathbun, 1893	P	mottled pea crab
Orthotheres strombi (Rathbun, 1905)	A	conch pea crab
Parapinnixa affinis Holmes, 1900	P	California Bay pea crab
Parapinnixa beaufortensis Rathbun, 1918	A	
Parapinnixa bouvieri Rathbun, 1918	A	
Parapinnixa hendersoni Rathbun, 1918	A	
Pinnaxodes floridensis Wells and Wells, 1961	A	polkadotted pea crab
Pinnixa barnharti Rathbun, 1918	P	
Pinnixa chacei Wass, 1955	A	Chace pea crab
Pinnixa chaetopterana Stimpson, 1860	A	tube pea crab
Pinnixa cristata Rathbun, 1900	A	
Pinnixa cylindrica (Say, 1818)	A	
Pinnixa eburna Wells, 1928	A	
Pinnixa faba (Dana, 1851)	P	mantle pea crab
Pinnixa floridana Rathbun, 1918	A	

SCIENTIFIC NAME	OCCURRENCE	COMMON NAME
Pinnixa franciscana Rathbun, 1918	P	
Pinnixa hiatus Rathbun, 1918	P	
Pinnixa leptosynaptae Wass, 1968	A	
Pinnixa littoralis Holmes, 1894	P	gaper pea crab
Pinnixa longipes (Lockington, 1876)	P	longleg pea crab
Pinnixa lunzi Glassell, 1937	A	Lunz pea crab
Pinnixa monodactyla (Say, 1818)	A	thumbless pea crab
Pinnixa occidentalis Rathbun, 1893	P	
Pinnixa pearsei Wass, 1955	A	
Pinnixa retinens Rathbun, 1918	A	
Pinnixa sayana Stimpson, 1860	A	
Pinnixa schmitti Rathbun, 1918	A	Schmitt pea crab
Pinnixa tubicola Holmes, 1894	P	
Pinnixa weymouthi Rathbun, 1918	P	
Pinnotheres chamae Roberts, 1975	A	jewel-box pea crab
Pinnotheres hemphilli Rathbun, 1918	A	
Pinnotheres maculatus Say, 1818	A	squatter pea crab
Pinnotheres moseri Rathbun, 1918	A	ascidian pea crab
Pinnotheres nudus Holmes, 1895	P	
Pinnotheres ostreum Say, 1817	A	oyster pea crab
Pinnotheres pugettensis Holmes, 1900	P	Puget pea crab
Pinnotheres shoemakeri Rathbun, 1918	A	
Pinnotheres taylori Rathbun, 1918	P	
Scleroplax granulata Rathbun, 1893	P	burrow pea crab

Superfamily Ocypodoidea

Ocypodidae—fiddler and ghost crabs

Ocypode quadrata (J. C. Fabricius, 1787)	A	Atlantic ghost crab
Uca burgersi Holthuis, 1967	A	saltpan fiddler
Uca crenulata (Lockington, 1877)	P	Mexican fiddler
Uca leptodactyla Rathbun, 1898	A	marbled fiddler
Uca longisignalis Salmon and Atsaides, 1968	A	gulf marsh fiddler
Uca marguerita Thurman, 1981	A	Olmec fiddler
Uca minax (Le Conte, 1855)	A	redjointed fiddler
Uca panacea Novak and Salmon, 1974	A	gulf sand fiddler
Uca pugilator (Bosc, 1802)	A	Atlantic sand fiddler
Uca pugnax (Smith, 1870)	A	Atlantic marsh fiddler
Uca rapax (Smith, 1870)	A	mudflat fiddler
Uca speciosa (Ives, 1891)	A	longfinger fiddler
Uca spinicarpa Rathbun, 1900	A	spined fiddler
Uca subcylindrica (Stimpson, 1851)	A	Laguna Madre fiddler
Uca thayeri Rathbun, 1900	A	mangrove fiddler
Uca vocator (Herbst, 1804)	A	hairback fiddler
Ucides cordatus (Linnaeus, 1763)	A	swamp ghost crab

Palicidae—stilt crabs

Palicus affinis A. Milne Edwards and Bouvier, 1899	A	Antillian stilt crab
Palicus alternatus Rathbun, 1897	A	labile stilt crab
Palicus cristatipes (A. Milne Edwards, 1880)	A	

SCIENTIFIC NAME	OCCURRENCE	COMMON NAME
Palicus cursor (A. Milne Edwards, 1880)	A	.... bathyal stilt crab
Palicus dentatus (A. Milne Edwards, 1880)	A	.... armored stilt crab
Palicus faxoni Rathbun, 1897	A	.... finned stilt crab
Palicus floridanus Rathbun, 1918	A	....
Palicus gracilis (Smith, 1883)	A	.... delicate stilt crab
Palicus obesus (A. Milne Edwards, 1880)	A	.... inflated stilt crab
Palicus sica (A. Milne Edwards, 1880)	A	.... winged stilt crab

Superfamily Cryptochiroidea

Cryptochiridae—gall crabs

Pseudocryptochirus hypostegus Shaw and Hopkins, 1977	A	....
Troglocarcinus corallicola Verrill, 1908	A	....

INDEX

Cambarus manningi (R. W. Bouchard)

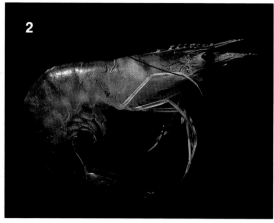

1 Sand snapping shrimp *Alpheus floridanus* (D. L. Felder)

2 White shrimp *Penaeus setiferus* (M. Vecchione)

3 Red brocade hermit *Dardanus insignis* (D. L. Felder)

4 *Callianassa louisianensis* (D. L. Felder)

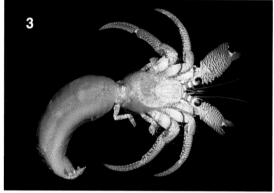

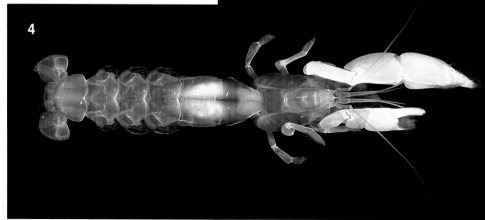

1 Spotted porcelain crab *Porcellana sayana* (D. L. Felder)

2 *Mithrax tortugae* (D. L. Felder)

3 Ridged slipper lobster *Scyllarides nodifer* (D. L. Felder)

4 Calico box crab *Hepatus epheliticus* (D. L. Felder)

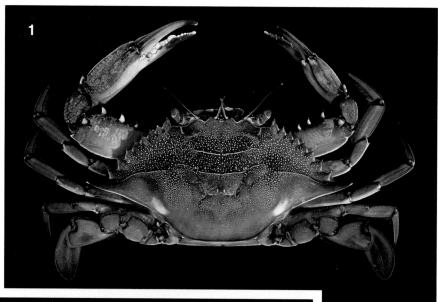

1 Blue crab *Callinectes sapidus* (D. L. Felder)

2 Atlantic sand fidler *Uca pugilator* (D. L. Felder)

3 Gulf stone crab *Menippe adina* (D. L. Felder)